T0311904

Mathematical Approaches to Liver Transplantation

Mathematical Approaches to Liver Transplantation

Eduardo Massad

Eleazar Chaib

Academic Press is an imprint of Elsevier
125 London Wall, London EC2Y 5AS, United Kingdom
525 B Street, Suite 1650, San Diego, CA 92101, United States
50 Hampshire Street, 5th Floor, Cambridge, MA 02139, United States
The Boulevard, Langford Lane, Kidlington, Oxford OX5 1GB, United Kingdom

Library of Congress Cataloging-in-Publication Data
A catalog record for this book is available from the Library of Congress

British Library Cataloguing-in-Publication Data
A catalogue record for this book is available from the British Library

ISBN: 978-0-12-817436-4

For information on all Academic Press publications visit our website at https://www.elsevier.com/books-and-journals

Publisher: Stacy Masucci
Senior Acquisitions Editor: Rafael E. Teixeira
Editorial Project Manager: Sara Pianavilla
Production Project Manager: Swapna Srinivasan
Cover Designer: Miles Hitchen

Typeset by TNQ Technologies

This book is dedicated to Professor Sir Roy Yorke Calne, pioneer of Liver Transplantation and introducer of cyclosporine, and Professor Silvano Atilio Raia, pioneer of Live-Donor Liver Transplantation.

Contents

About the authors

Eduardo Massad is Professor Emeritus of Medical Informatics of the School of Medicine, University of Sao Paulo, Brazil, and Professor of Applied Mathematics of the Fundacao Getulio Vargas, Rio de Janeiro, Brazil. He is Fellow of the Institute of Mathematics and its Application and Fellow of the Royal Society of Medicine, UK.

Eleazar Chaib is Associate Professor of Transplantation Surgery of the School of Medicine, University of Sao Paulo, Brazil, and is Fellow of the Royal College of Surgeons, UK.

Foreword

Liver transplantation is a spectacular medical success story: it is the only effective therapy for large numbers of patients with end-stage liver disease, metabolic liver disease, and primary liver cancer. There have been enormous strides in the success of the procedure: 1-year patient survival rates better than 90% are routinely achieved, with 5-year survival rates in excess of 70% in the best centers.

However, the huge achievement of liver transplantation over the last 50 years is matched by the massive challenge of meeting the growth in demand. Liver transplantation has become a victim of its own success, compounded by rapidly increasing rates of liver disease around the world. Despite increasing organ donation rates (in some countries), waiting lists for liver transplantation continue to increase: patients are more likely to die on the waiting list than in the 12 months postoperatively.

Indeed, the problem is greater than is immediately apparent: in order to direct scarce donor organs to those patients most likely to benefit, many patients with poorer prognoses are not listed for transplantation. Better access to transplantation would improve the survival of many more patients, particularly those suffering from cancer confined to the liver. While some causes of chronic liver disease, such as hepatitis C, will become less prevalent, others are emerging: nonalcoholic steatohepatitis is a rapidly increasing cause of end-stage liver disease.

Innovative approaches that unlock a wider spectrum of donor organs (including those previously thought unsuitable) have increased transplant rates in many countries but not sufficiently to alter the underlying trend of an increasing discrepancy between demand and supply. The development of successful living donor liver transplantation programs has gone some way to alleviate the problem, particularly in countries where deceased donor transplantation is less available, but suitable donors are not available to most patients.

Organ allocation has become a topic of intense discussion and scrutiny because of the need to make judgments as to which patients receive this critically scarce resource, knowing that patients who do not receive priority may not survive. Competing factors within this debate include the need to maximize the benefit from available donors (utility), the need to prioritize those patients who cannot wait (urgency), and the desire to treat all patients equitably irrespective of prognosis (fairness).

The demands of liver transplantation are not only technical but also logistic and ethical. The success of this method of treating patients with end-stage liver disease, first carried out successfully in 1967, is extraordinary, but the challenges posed to the current generation of transplant specialists, although different, remain as great as ever. In this book, Eleazar Chaib and Eduardo Massad uniquely combine clinical narrative and mathematical modeling, to depict the practice, together with the current and projected challenges of liver transplantation, with particular reference to its evolution in Sao Paolo, Brazil. This monograph foregrounds the challenges, benefits, and paradoxes of liver transplantation.

Peter Friend MD, FRCS
Professor of Transplantation
University of Oxford Medical Sciences Division
John Radcliffe Hospital
Headington, Oxford, United Kingdom

Preface

We first met in the end of 1973—virtually—in the list of approved students for entrance in the Santa Casa de São Paulo Medical School; EC in 23rd place, EM in 24th. One month later, we met again—virtually—in the list of approved students for entrance at the very prestigious School of Medicine of the University of São Paulo; EM 21st, EC 22nd. Finally, we met in person at the first day of class at the medical course. It was the beginning of a friendship that only strengths with time (EC happens to be the Godfather of EM's youngest daughter!).

Although keeping a close friendship, our professional careers diverged away and EC followed a path as a transplantation surgeon and EM as a theoretical and mathematical biologist. After two stints each in the UK (EM with Professor John Maynard Smith at Sussex University and Professor Sir Roy Anderson at the Imperial College of Science Technology and Medicine in the early 1980s; EC with Professor Sir Roy Calne at Cambridge University and Professor Peter Friend at Oxford University at the early 1990s and 2000s), we decided, some years ago, to attempt a marriage between our expertise, which sounded very weird at the time: mathematics of liver transplantation! Our first publication in this field appeared in 2005 and it inaugurated a new approach to the issue of liver transplantation; backing off the relatively narrow surgical field (EC cup of tea) to a bird's eye of the population dynamics (EM cup of tea).

Many publications later, we sat at the Costa Café in the basement of Waterstone bookshop of Gower Street to combine all our past experience in the form of a book (we regret to say that there is no Costa Café anymore at Waterstone!).

EM first suggested the title "The Mathematics of Liver Transplantation." EC rebuked that this title would scare away the surgeons. EM agreed and suggested "The Population Biology of Liver Transplantation," a more edible alternative for EC appetite, but at the same time, a slightly misleading title. After consulting Sir Roy Calne, who kindly agreed to write down the Foreword, we settled on the title "Mathematical Approaches to Liver Transplantation."

We wrote this book not only with the traditional transplantation surgeon in mind, who could indeed profit from this new approach, but in particular, for decision makers that, through the quantitative methods here presented, could optimize the distribution of the frequently limited number of liver grafts, in order to reach the most just programme of transplantation organs distribution.

We hope any reader interested in either Biomathematics or Transplantation Surgery (or both, an unlikely rare bird!) can profit from this work. The mathematical and surgical technicalities are kept to a minimum and the more interested reader in either field (or both) can go deeply in the material presented in the appendices.

We are deeply in debt to a wide range of friends and colleagues who participated in this bold adventure to unchartered seas: ...

Eduardo Massad
Eleazar Chaib

Acknowledgments

Several colleagues contributed to this work, either by coauthoring the papers that generated this book or by providing advice on many aspects of the text. First and foremost, we would like to thank Professor Luiz Augusto Carneiro D'Albuquerque, Director of the Liver Transplantation Unit of the Clinics Hospital of the School of Medicine of the University of São Paulo, for providing the conditions for carrying out the research that basis this book. We are also thankful to Drs. Francisco Antonio Bezerra Coutinho, senior mathematician on many papers related to this book, Marcelo Nascimento Burattini, infectious diseases clinicians who provided many clinical advice on some of the chapters, and Marcos Amaku for helping with many of the mathematical hurdles related to some of the chapters and lastly, but not least, to our families for the emotional support to the development of this work.

Introduction

Over the last 50 years, organ transplantation has achieved great success to become standard therapy for the treatment of patients with end-stage organ failure. The most frequent indications for liver transplantation are alcoholic liver disease, hepatocellular carcinoma, and viral hepatitis. Notwithstanding this success, it has emerged candidate waiting lists that greatly outnumber the current supply of deceased donor organs. The increasing number of candidates and transplants performed has resulted in demand for deceased donor liver transplantation (DDLT) that vastly exceeds the supply. In the United States only (2003), there are over 17,000 patients listed for liver transplantation in the UNOS database, yet only 5600 deceased donor grafts became available. This disparity between supply and demand has widened every year since 1994, thus necessitating a system to prioritize the vast number of patients waiting on a limited pool of donor organs. Over the years there have been several fundamental changes in the prioritization of patients for liver transplantation.

Also in Europe the liver allocation program is complex because allocation rules need to follow not only the guidelines of the European Commission but also the specific regulations of each of the seven Eurotransplant countries with active liver transplant programs. Thirty-eight liver transplant centers, approximately, served a population of about 135 million in 2015. Around 1600 DDLT are transplanted annually. The number of deceased organ donors remains stable, but donor age is increasing. Nevertheless, liver utilization rates are unchanged at around 80%.

The increase in the number of donors is the main objective of all transplantation organizations around the world. Further understanding of the factors involved in increasing donation rates is very important for planning future strategies to improve outcomes in each country.

Fig. 1.1 shows the number of deceased organ donors per million population (pmp) according to the countries; for example, Spain has the highest number of donors of 35 and United States has 20 donors per million of population, respectively.

Mathematical Approaches to Liver Transplantation. https://doi.org/10.1016/B978-0-12-817436-4.00001-1

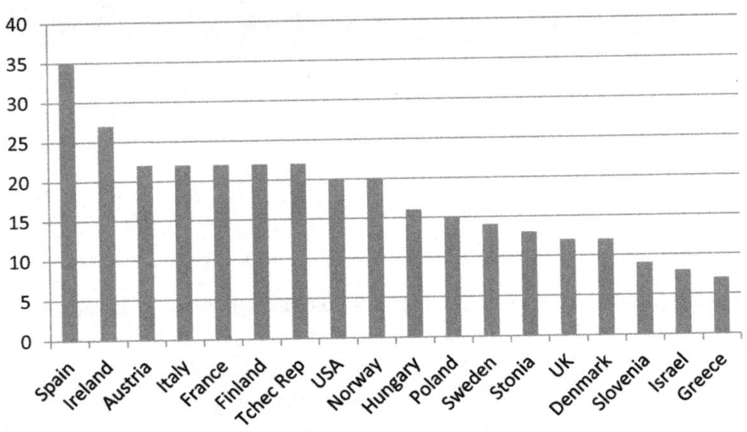

■ **FIGURE 1.1** Number of deceased organ donated per million population according to the countries.

In addition to that, in the United Kingdom only in the decade of 2003–12, we have seen major changes in how organ donation and transplantation is delivered. The number of DDLT has increased from 709 (12.0 pmp) to 1164 (18.3 pmp); this increase has been predominantly a result of an increase in donors after circulatory death (from 1.1 to 7.9 pmp) while the number of donors after brain death has remained broadly stable (around 10.5 pmp).

To select deceased donors, the transplant community worldwide first used the Child-Turcotte-Pugh (CTP) score to create an objective scoring system. In addition, the score OPTN/UNOS (USA) has been adopted in 1998 because of its predictive ability regarding candidates with complications of portal hypertension. The score, unfortunately, lacked the ability to discriminate severity of illness for using subjective variables and more was not validated in candidates on the liver transplant waiting list.

The MELD (model for end-stage liver disease) score was originally developed as a prognostic tool in cirrhotic candidates undergoing transjugular intrahepatic portosystemic shunts. The MELD score was validated for patients on the liver transplant waiting list both because it is more accurate than CTP and because it has better prediction of 3-month waitlist mortality. Since 2002, DDLT allocation in the United States has been based on MELD and pediatric end-stage liver disease scores. This urgency-based system prioritizes candidates by mortality risk while they await liver transplant and is recognized as a major improvement over previous allocation policy. The MELD score has shortened the DDLT waiting list and improved waiting list survival without adversely affecting posttransplant survival.

Our book has the proposal of discussing ways of improving donor allocation both in donors after circulatory death and donors after brain death using mathematical models applied in liver cirrhosis with and without liver tumor. We back off from the classical surgical view, which centers on the individual patients, and assume a bird's view of liver transplantation as a population issue. We aim to attenuate the disproportion between the numbers of available liver grafts versus waiting list recipients by proposing methods to optimize the donor grafts distribution. The discrepancy between the number of patients on the waiting list and available donors remains the key issue and is responsible for the high rate of waiting list mortality. By using sophisticated mathematical tools, we check, from a population biology perspective, the current and future strategies to the best use of limited number of organs (Pérez and Castroagudin, 2010).

We honestly hope that the methods presented in this book may be helpful in reducing the suffering and mortality of liver failure patients desperately waiting on interminable lists for a new liver.

1.1 TYPOLOGICAL REASONING VERSUS POPULATION REASONING

From the essences of Plato to the Linnaeus taxonomy, typological reasoning dominated biological thinking until the early 19th century (Mayr, 1982). According to this school of thought, the type (or essence) was the truth and all variability should be abstracted and disregarded. This form of reasoning permeated other areas of human knowledge beyond biology and an example that persists to this day is astrology. This pseudoscience occupied the time of several thinkers such as Galileo, Kepler, and Newton, just to name a few. According to the characteristically typological reasoning of astrology, human beings are divided into 12 classes, or types, according to the sign of the zodiac to which they belong. The whole life of a person, therefore, is determined by the date of his/her birth and by the movement of some (few) celestial bodies. A person of the Libra sign has certain characteristics and is expected to behave according to be predicted by his sign. Any variation around the expected type is abstracted as irrelevant. Another example very characteristic of typological reasoning and that still persisting until today is the one of the homeopathy. According to this school of thought, people are divided into classes or types and all their diseases are caused by certain "energy imbalances," whose restoration is made by quasi-atomic doses of the elements that characterize their type. Thus, for example, a person of the Natrium muriaticum type should be treated, regardless of their disease (as a general guideline), with cooking salt in nanoscopic dilutions. With the perception that nature is highly heterogeneous and that our ability to control certain natural phenomena depends on our ability to deal with variability, the population school of thought was born at some point in the 19th century. Certainly, Victorian biologists played a decisive role in this conceptual revolution. According to population reasoning, type becomes an abstraction of reality, while variability becomes the very expression of reality. Thus, for example, in current language, when we say that the average number of patients with heart disease who are attended per day in a certain emergency care service is 37.5 individuals, the individual mean is obviously an abstraction. Likewise, when we say that the mean systolic blood pressure of a certain patient group is 125 mm Hg, it may happen that no individual in this sample has 125 mm Hg. The average is, therefore, an abstraction of reality and should be regarded as a representative of that population. The variability, expressed in current language as the variance, is reality. According to Ernst Mayr (1982), one of the greatest living biologists, the replacement of typological reasoning by the population represents the greatest conceptual revolution in all biology. Without it, we could not deal with variability, nor could we reason in terms of evidence, genomics, and much less in evolutionary medicine.

1.2 GENERAL INTRODUCTION TO MODELS THEORY IN BIOLOGY

In a series of works published in the period from 1637 to 1649, Descartes introduced the concept of reductionism, understood as the study of nature as composition of its physical components, which can be analyzed separately. Reductionism and analytical mathematical models have since been destined to become the most powerful intellectual tool of modern science (Wilson, 1998). The design of mathematical models of complex (and often nonlinear) real systems is therefore essential in many branches of science (Lindskog and Ljung, 1997). The proposed and developed models can be used, for example, to explain the underlying behavior of real systems and to serve as instruments of prediction and control. A common approach in the construction of mathematical models is the so-called black box modeling (Ljung, 1987; Söderstrom and Stoica, 1989), as opposed to more structured modeling, based on so-called "first principles" of physicists, also known as white box, where everything in the system is considered a priori from the "physical" (also known in biology as Natural History) of the real system. A black box model is architected entirely from data, without considering any system physics. The structure of the model (not the system) is chosen from families of known models with flexibility and successful earlier applications. This implies that the parameters of the model (see below) also lack physical or even verbal meaning; these are chosen only to fit the observed data to the best possible extent. This class of models is also called a posteriori, meaning that the model is constructed after the data are collected. Examples of this class of models are the adjustments of data through statistical regression techniques. Contrasting with the black box models, the white box models are intended to mimic the behavior of real systems. Thus, the natural history of the phenomenon to be studied is incorporated into the model and, therefore, these models are known as structured models; that is, the structure of the system is taken into account. From the mathematical point of view, this translates into the form of systems of differential equations (or difference), to reproduce, in the best possible way, the dynamics of the system. Also known as a priori, models may have considerable predictive capacity and control utility. To construct a mathematical model, several decisions are necessary, both explicit and (more often) implicit (Gershenfeld, 1999). There is no formal mechanism to guide the choice or even an acceptable definition as to which model is best. A good attempt is Rissanen's (1986) proposal of the so-called minimum description principle, essentially a version of Occam's razor: the best model is the one that has the smallest size, including information that specifies both the shape of the parameters and how to communicate the results. In general, however, the best model, according to Gershenfeld (1998), is the one that works best for you.

History of liver transplantation

Medawar, a biologist from Oxford, noted that allografts were a practical means of covering burn defects based on Gibson's work of pinch grafts from allogenic sources. He also noted that their dissolution was accelerated at a second exposure. The term "second set," which ascribes memory to the immune system, dates to this work and has become an important aspect of immunologic research. Next, he became interested in recipients of progressively young ages. Gibson showed that mice exposed to allogenic cells while in utero became tolerant to transplanted tissue even into adult life (Billingham et al., 1953).

Based on this principle, Hume at Peter Bent Brigham Hospital in Boston tried to modify the immune response to allografts in human performing renal transplantation using unrelated donors and giving no immunosuppressive therapy combined with total body irradiation and bone marrow transplantation (Busuttil and Klintmalm, 2015). These efforts almost uniformly ended in failure.

Schwartz and Deneshek (1959) showed that 6-mercaptopurine prevented an antibody response in rabbits given a foreign protein. This was the first drug shown to produce persistent immunologic tolerance.

Calne working in Francis Moore's laboratory tried this drug in canine renal experiments and found, for the first time, that the immune system can be controlled at least in some instances (Calne, 1960). In 1960, 1 year after its description in Nature, this drug was used in human kidney transplant at the Peter Bent Brigham Hospital (Murray, Merril, and Dammin, 1063). For the first time, it became apparent that a degree of tolerance could be achieved in humans by using pharmacologic immunosuppression.

Starzl, 2 years later, showed that azathioprine and prednisone combination was effective as prophylaxis against rejection in kidney transplantation in humans (Starzl et al., 1963).

Mathematical Approaches to Liver Transplantation. https://doi.org/10.1016/B978-0-12-817436-4.00002-3

The first liver transplant in humans was attempted by Starzl in 1963. Long-term survival was not attained in his first five cases. After overcome and solved both technical and cold ischemia problems, in July 1967, a liver of a child was transplanted. The child survived 13 months before dying of recurrent hepatoma.

In spite of the technical aspects of the liver transplantation defined, such as bile duct reconstruction, coagulation support, and refinement of donor procedure (including the acceptance of the concept of brain death in 1968), the first-year mortality was high approximately 70% (Starzl et al., 1963).

Another pioneer Calne added important laboratory observations and technical modifications to the transplant procedure (Calne and Willians, 1979; Calne, 1983). Moore visiting Calne in Cambridge assisted what is probably the first hepatic transplant outside America.

The most important event in the field of immunosuppression in transplantation was also reported by (Calne et al., 1978), the introduction of cyclosporine. The earliest clinical study of the use of cyclosporine in transplantation was conducted by Calne and his group demonstrating effective immunosuppression in liver, pancreas, and renal transplantation, although with an alarming incidence of significant toxic side effects, including nephrotoxicity, infection, and development of lymphoma (Calne et al., 1979).

Calne and his group were also the first to use cyclosporine in clinical hepatic transplantation, but it was Starzl and his group who first showed its beneficial effect on postoperative survival when used in combination with steroids (Starzl et al., 1981). With the availability of cyclosporine-A, 1-year survival immediately rose from 30% to approximately 65%—75%. This was the event that made the procedure practical.

REFERENCE

Calne, R.Y., White, D.J., Thiru, S., Evans, D.B., McMaster, P., Dunn, D.C., Craddock, G.N., Pentlow, B.D., Rolles, K., 1978. Cyclosporin A in patients receiving renal allografts from cadaver donors, Lancet. Lancet 2 (8104—5), 1323—1327.

Causes of liver failure

Today, the possibility of liver transplantation is considered in almost every patient presenting with evidence of liver failure. The liver diseases for which liver transplantation has been performed in adults can be divided into three broad categories of advanced chronic liver disease, fulminant hepatic failure, and unresectable hepatic malignancy.

Other small category of diseases for which liver transplantation may be indicated is inherited metabolic liver diseases in which inborn error of metabolism resides in the hepatocyte, such that liver transplantation is curative.

3.1 CHRONIC HEPATIC FAILURE

This category includes the following conditions:

1. Primary biliary cirrhosis
2. Chronic active hepatitis
   - **2.1.** Autoimmune
   - **2.2.** Posthepatitis
3. Primary sclerosing cholangitis
4. Biliary atresia
5. Budd–Chiari syndrome
6. Cryptogenic cirrhosis
7. Alcoholic cirrhosis

In these patients with terminal parenchymal liver disease, the decision of how and when to perform liver transplantation may be difficult. As a general rule, the transplant is indicated under the following conditions.

3.2 DEVELOPMENT OF SITUATIONS THAT RISK THE PATIENT'S LIFE

1. Recurrent hemorrhagic esophageal varices
2. Encephalopathy

Mathematical Approaches to Liver Transplantation. https://doi.org/10.1016/B978-0-12-817436-4.00003-5

3. Spontaneous bacterial peritonitis
4. Intractable ascites

Patients with alcoholic cirrhosis have been submitted, to a lesser extent, to liver transplantation because of the high alcohol recidivism rate, approximately 30%. Potential candidates will be those in whom there is good evidence of alcohol withdrawal for many months (at least 6 months) but also have a satisfactory psychiatric evaluation score of 12 and 14.

3.3 FACTORS TO BE CONSIDERED IN THE EVOLUTION OF PATIENTS WITH ALCOHOL-INDUCED LIVER DISEASE LISTED FOR LIVER TRANSPLANTATION

Medical

1. Extension and potential reversibility of liver injury, presence of
   - 1.1. Jaundice
   - 1.2. Hypertension portal
   - 1.3. Ascites
   - 1.4. Encephalopathy
2. Evaluation of alcohol-related injuries in other organs that may limit survival after liver transplantation
   - 2.1. Heart
   - 2.2. Pancreas
   - 2.3. Peripheral nervous system
   - 2.4. Central nervous system
3. Nutritional status
   - 3.1. Protein deficiency
   - 3.2. Vitamin deficiency
4. Presence of infections
   - 4.1. Tuberculosis
   - 4.2. Concomitant viral hepatitis (B, C, D)

Social

1. Possibility of returning to alcoholism (motivation, family, and institutional support)
2. Availability of organs

3.4 DETERIORATING THE QUALITY OF LIFE

Liver transplantation is indicated when the activities of patients are so restricted by their disease that they become incompatible with an acceptable lifestyle.

3.5 STOP OF GROWTH (IN CHILDREN)

In children, liver transplantation is delayed as much as possible so that they can reach their development before the operation. This reduces surgical, anesthetic, and postoperative complications. When the disease is advanced to the point of paralyzing its growth, the transplantation is performed without delay.

3.6 ACUTE HEPATIC FAILURE

This category covers two groups of patients:

1. **Fulminant hepatic failure** (onset of encephalopathy up to 8 weeks after onset of symptoms)
2. **Subacute liver failure** (onset of encephalopathy between 8 and 26 weeks after onset of symptoms)

The main causes of acute liver failure are

1. Viral hepatitis (A, B, C, D, E); herpes simplex; Epstein–Barr; *Cytomegalovirus*
2. Drug reactions (acetaminophen, halothane, antituberculin, antiepileptic)
3. Steatotic liver
4. Budd–Chiari syndrome
5. Acute Wilson's disease
6. Sepsis
7. Tumor (malignant infiltration, Hodgkin's lymphoma, and non-Hodgkin's lymphoma
8. Veno-occlusive disease
9. Hepatic vein thrombosis
10. Ischemia and hypotension

This category of patients has benefited from liver transplantation only recently, with an increase in the number of donors. In the past, the probability of obtaining an organ within a short period of time between the onset of conditions that would indicate the transplantation and the recipient was so small that most of the patients were not transplanted.

Transplantation is indicated in patients who worsen clinically laboratory, with changes in liver function tests, coagulopathy, and encephalopathy. Early referral of these patients to referral centers for transplantation is advisable, as their clinical-laboratory deterioration is so rapid that transplantation becomes a hazard, but usually they are hypotensive, anuric, and have elevated intracranial pressure.

3.7 CONGENITAL ERRORS OF METABOLISM

This category includes patients who have hepatic enzymatic defects that put their lives at risk.

They are divided into two categories:

Conditions in which the liver is the target organ and is primarily affected

1. Alpha 1 antitrypsin deficiency
2. Wilson's disease
3. Glycogen storage disease
4. Protoporphyria
5. Crigler—Najjar Syndrome
6. Galactosemia

Conditions in which an extrahepatic organ is primarily affected

1. Primary hyperoxaluria
2. Primary hypercholesterolemia

This group of patients, many of whom are children, can be cured when replacing the liver producing the deficient enzyme. In some cases, the injury in another target organ may need a transplant for this second organ reached. Many patients with renal failure secondary to hyperoxaluria have been successfully treated by the associated kidney and liver transplants. In many of these conditions, hepatic function is normal in all its aspects except for enzyme deficiency (e.g., hyperoxaluria). In others, the lesion structural changes in liver result in the development of cirrhosis and complications of chronic liver failure (e.g., alpha 1 antitrypsin deficiency, Wilson's disease).

The timing for the indication of transplantation is based on the development of the complications of each particular disease.

3.8 TUMORS OF THE LIVER

Malignant tumors of the liver represent a significant, but declining, proportion of patients undergoing liver transplantation. At the beginning of the development of liver transplants, malignant tumors were considered a good indication for transplantation, as such patients did not have the paraquenchimatous components of liver disease such as portal hypertension and coagulopathy and were supposedly more likely to survive the operation. However, the long-term results of transplantation in these patients showed an enormous incidence of tumor recurrences.

The patients in this group are distributed among the following diagnoses:

Primary hepatoma
Cholangiocarcinoma
Other primary liver tumors
Secondary tumors

The results of liver transplantation for secondary liver tumors are very poor, even in cases where the primary tumor was removed several years ago. Therefore, these cases are not currently on the priority list of transplants, except in very special circumstances such as symptomatic recurrent carcinoid tumors.

The outcome of liver transplantation for cholangiocarcinomas has also shown disastrous results, with recurrence rates greater than 50% at 12 months. Even patients with signs of extrahepatic tumor invasion are not found at time of operation. The recurrence of tumors tends to progress very rapidly in the presence of immunosuppressive medication.

The best reported results are in patients with primary hepatoma, especially in fibronuclear tumors and in certain rare primary tumors of low malignancy.

Liver transplantation may be indicated in primary liver tumors under the following conditions:

1. The tumor should not be resectable by conventional liver surgery
2. There should be no sign of extrahepatic tumor disease

In patients in whom the transplant decision is made, the operation should be performed as soon as possible.

3.9 CLINICAL AND BIOCHEMICAL INDICATIONS IN LIVER TRANSPLANT CANDIDATES

Acute hepatic failure

1. Bilirubin $>10-20$ mg/dL, with a tendency to increase
2. Prothrombin time >10 s above the control, with a tendency to increase
3. Progressive hepatic encephalopathy (at least grade 3)

Chronic hepatic disease

Hepatic colestatic disease

Bilirubin $>10-15$ mg/dL
Intractable pruritus
Intractable bone disease
Malnutrition
Recurrent cholangitis

Hepatocellular hepatic disease

Albumin <2.5 g/dL
Hepatic encephalopathy
Prothrombin time >5 s above the control, with a tendency to increase

Factors common to both types: colestatic and hepatocellular

1. Portal hypertension with bleeding (esophageal varices)
2. Hepatorenal syndrome
3. Recurrent spontaneous bacterial peritonitis
4. Intractable ascites
5. Recurrent bile duct sepsis
6. Appearance of hepatocellular carcinoma

Clinical indications for emergency liver transplantation

1. Uncontrollable hemorrhagic esophageal varices
2. Acute renal failure due to hepatorenal syndrome
3. Hepatic encephalopathy Grade III (delirium, drowsiness) and IV (coma)

Contraindications to liver transplantation

Absolute

1. Sepsis outside the hepatobiliary system
2. Metastatic hepatobiliary tumors
3. Immunodeficiency syndrome
4. Severe pulmonary hypertension
5. Advanced cardiopulmonary disease
6. Alcoholism present

Relative

1. Age over 65 years
2. Advanced chronic kidney disease
3. Portal vein thrombosis
4. Cholangiocarcinoma
5. Hypoxia with shunt from right to left
6. Hepatitis B with AgHbe positive
7. Previous shunt cava portal
8. Prehepatic biliary surgery
9. HIV+ without clinical symptoms of AIDS
10. Diabetes mellitus
11. Psychiatric illness in the past

Technical and surgical aspects of liver transplantation

4.1 DONOR OPERATION

The general criteria for choosing a suitable liver donor are as follows:

1. Donor in mechanical ventilation with beating heart
2. Diagnosis of brain death
3. Family consent
4. Stable circulation
5. Age of 65 years
6. No history of liver disease, drug abuse, alcohol abuse, malignant tumors, and sepsis
7. AgHbs and HIV negatives

4.2 EVALUATION OF POTENTIAL LIVER DONOR

Clinical history

1. Cause of death
2. Cardiorespiratory arrest (number and duration)
3. Systemic hemodynamic stability: urinary output, blood pressure, concentration of inotropic drugs
4. Associated injuries or prior operations
5. Previous infections and antibiotics
6. Time interval from hospital admission
7. Nutritional status
8. Documentation of brain death
9. Disseminated intravascular coagulation
10. No tumor outside the central nervous system
11. Absence of severe hepatic trauma

Mathematical Approaches to Liver Transplantation. https://doi.org/10.1016/B978-0-12-817436-4.00004-7

12. Absence of systemic sepsis
13. No prolonged hypoxia

Laboratory tests

1. Serology of hepatitis B and C virus
2. Serology for syphilis (VDRL), AIDS (HIV), and cytomegalovirus (CMV)
3. ABO blood type
4. Total bilirubin, transaminases (TGO, TGP normal, or near normal)
5. Urea and creatinine, Na and K
6. Complete blood count
7. Coagulogram (TP, TTPA, platelets)
8. Gasimetry and fraction of inspired oxygen

4.3 THE DONOR'S HEPATECTOMY

This operation is usually associated with removal of the kidneys, heart, and eventually lungs.

Preparation

The donor, with diagnosed encephalic death with device-controlled breathing, is positioned in the dorsal decubitus position, placing the operative fields so as to allow extensive exposure of the cervical region, thorax, and abdomen. The antibiotics used intravenously and in the infusion liquid are as follows:

1. Prehepatectomy: Imipenem 500 mg EV
2. In the infusion liquid: Ampicillin 1 g/L and gentamycin 8 mg/L

Incision

The incision and median extending from the sternal furcula to the pubic symphysis are shown in Fig. 4.1.

Operative technique

1. Ligature of the round ligament with retrograde dissection of the triangular ligament to the height of the suprahepatic vena cava.
2. Evaluation of the arterial supply of the liver (normal irrigation and anatomical variations, Figs. 4.2 and 4.3). The elements of the hepatoduodenal ligament in the minor curvature are dissected, and the bile duct is taken down very close to the duodenum.

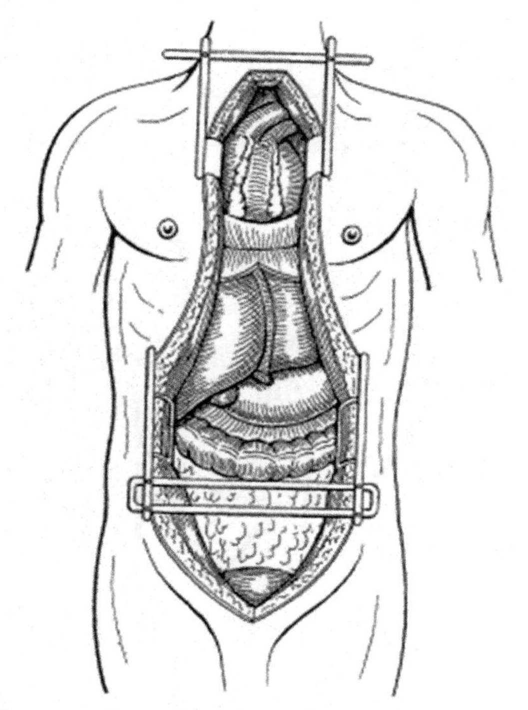

■ **FIGURE 4.1** Donor surgery: Sternum abdominal incision for exposure of organs of the thoracoabdominal cavity.

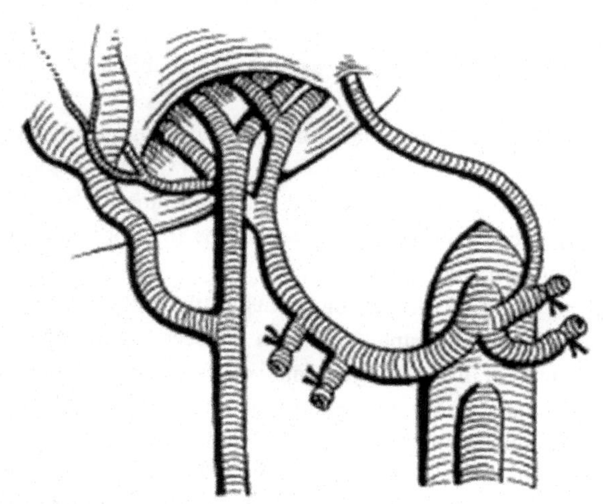

■ **FIGURE 4.2** Donor surgery: Left hepatic artery originating from the left gastric artery (anatomical variation).

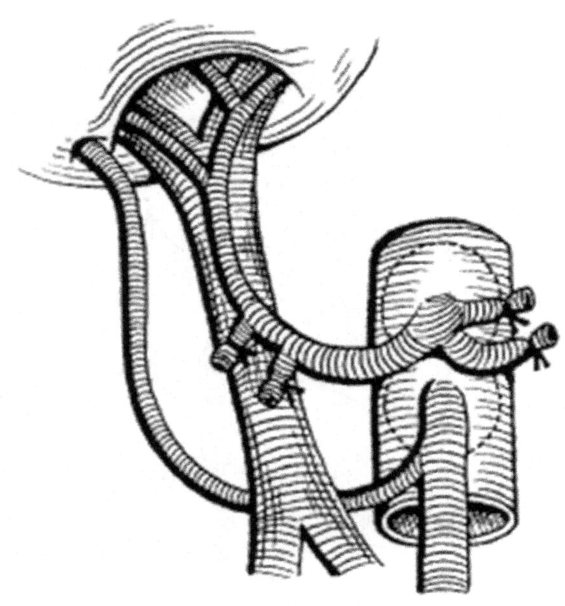

■ **FIGURE 4.3** Donor surgery: Hepatic artery originating from the superior mesenteric artery (anatomic variation). In this case, the Carrel patch of the aorta is removed, encompassing the emergence of the celiac trunk and the superior mesenteric artery.

3. Ligation of the gastroduodenal, splenic, and left gastric arteries and identification of the celiac trunk. If there is a hepatic artery (left hepatic artery originating from the left gastric or right hepatic artery from the superior mesenteric artery), they are carefully dissected. Thus, all arteries may be perfused by the cannula in the abdominal aorta Figs. 4.2 and 4.3.
4. Ligature of all small branches that enter the portal vein (left gastric vein, etc.).
5. Mobilization of the right colon and small intestine for exposure to the inferior vena cava and abdominal aorta. The inferior vena cava and renal veins are repaired with tape.
6. Dissection of the peritoneal ligaments of the left and right lobes of the liver including the small omentum. Dissection of the hepatic nude area and ligation of the right adrenal vein, releasing the liver and the inferior vena cava from the posterior abdominal wall.
7. Patient heparinization (20,000 IU EV) and placement of infusion cannulas in the distal aorta and portal vein, via superior and inferior mesenteric veins. The inferior vena cava may be used for drainage of

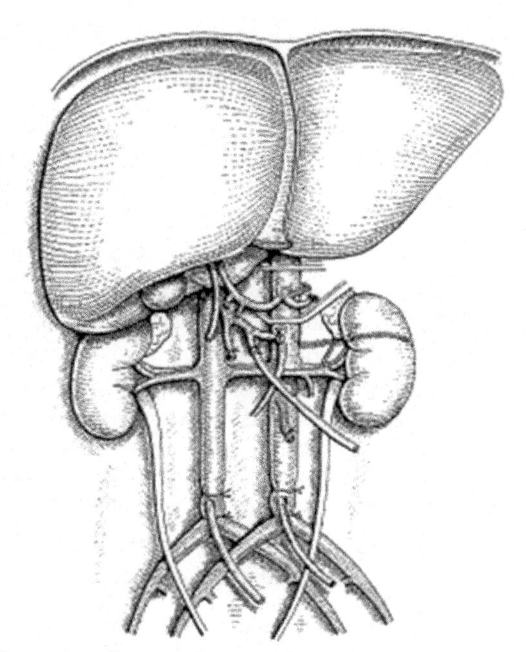

■ **FIGURE 4.4** Donor surgery: Placement of hepatic perfusion cannulas in the portal vein and abdominal aorta and collection of fluid in the inferior vena cava.

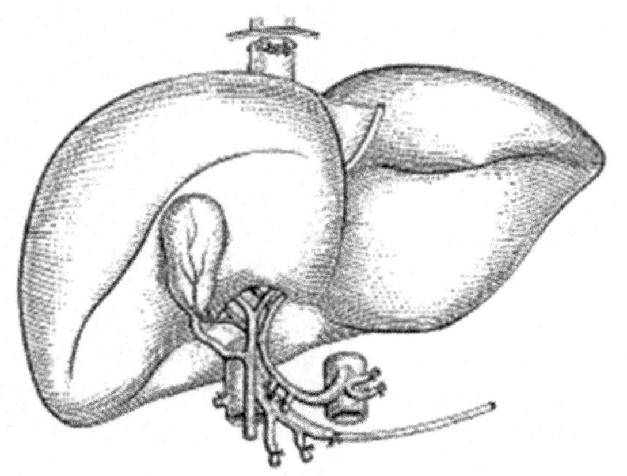

■ **FIGURE 4.5** Donor surgery: Placement of the hepatic perfusion cannula in the splenic vein (technical variation).

hepatic perfusion fluid (Fig. 4.4). Alternatively, a hepatic lavage perfusion probe may be placed in splenic vein instead of the portal vein (Fig. 4.5).

8. Immediately before the cardiotomy, clamp the aorta in the diaphragmatic hiatus and perfuse the liver with the preservation solution via the portal vein and hepatic artery. The blood and preservation fluid are drained from the pericardial cavity by the suprahepatic inferior vena cava section. The liver can also be drained through the infrahepatic inferior vena cava through a cannula placed inside it.
9. When the infusion is completed, the portal vein and the inferior vena cava are sectioned. The arterial supply is removed with a patch of the aorta and liver removed from the abdominal cavity.
10. In the so-called "back table" works, the three phrenic veins are sutured into the suprahepatic inferior vena cava, and the gallbladder, common bile duct, and hepatic artery are infused with hepatic perfusion solution (+200 mL).
11. The liver is then placed in a sterile bag with the preservation solution and ice wrapped to be transferred to the transplant center.

In urgent cases, it is possible to transplant a wolf or an adult liver segment. This has been used in a good number of children who would be unable to survive long enough to wait for a suitable organ to become available.

The donor organ preservation time (cold ischemia) was only recently reduced from 8 to 12 h. For this reason, many centers have two surgical teams so that the recipient's surgery can begin before the donor's team returns, thus decreasing the time of cold ischemia.

However, with the development of new preservation solutions, it is possible to maintain the liver in a container with ice (4°C) for up to 24 h. This has altered the logistics of liver transplantation allowing the recipient's surgery to be done by appointment, if necessary, a few hours after that of the donor. This also made the availability of organs, especially those coming from long distances to the transplantation centers, widen.

In the table below, one can compare the chemical and electrolytic composition of some of the liver preservation solutions used in the past (Ringer's lactate) and in the present (Euro-Collins, Cambridge II, and UW) Fig. 4.4, Fig. 4.5, Fig. 4.6, Fig. 4.7, Fig. 4.8, Fig. 4.10.

4.4 COMPOSITION OF HEPACTIC PRESERVATION SOLUTIONS

	Ringer's lactate	Euro-collins	University of Wisconsin	Cambridge II
Electrolytes (mM/L)				
Ca	1,5			
Cl	109	15		
Mg			5	5
Na	130	10	30	120
K	4	115	120	30
Anions (mM/L)				
HCO_3		10		
Lactate	28			
Lactobionate			100	100
Phosphate		57.5	25	
Colloids and impermeants (g/L)				
Hidroxyethyl			50	
Raffinose			17.8	30
Others				
Glucose (g/L)		194		
Adenosine (mM/L)			5	
Allopurinol (mM/L)			1	1
Glutathione (mM/L)			3	3
Insulin (μ/L)			100	
Antibiotics and steroids				
Sulfamethoxazole (mg/L)			40	
Trimetroprim (mg/L)			8	
Dexometasone (mg/L)			8	
Ampicillin (g)				1
Gentamicin (mg)				8
Osmolarity	**273**	**375**	**325**	**305–315**
PH	**6.3**	**7.4**	**7.4**	**7.4**

4.5 THE RECIPIENT OPERATION

The specific criteria for choosing the recipients are

1. Blood group compatible with donor
2. Compatibility in liver size (especially in children)

Surgery at the liver transplant recipient is considered to be large in both surgical and anesthetic terms. It requires not only a very well-trained surgical staff but also great laboratory support. This procedure rarely requires great support from the blood bank as well as biochemistry laboratory. The results of these analyzes should be provided regularly and as quickly as possible.

Liver mobilization is often complicated by factors such as

1. Portal hypertension
2. Coagulopathy
3. Vascular adhesions secondary to previous abdominal surgery

Incision

The Mercedez incision is the medial incision combination below the xiphoid appendix with a supraumbilical transverse incision on both sides of the abdomen.

Operating techniques

1. Suture the anterior parenchyma of the abdomen to bring close to the nipples and position the retractable retractor by pulling up the costal borders bilaterally toward the patient's head,
2. Dissection of the vena cava above and below the liver with release of the peritoneal ligaments of the liver and ligation of the right adrenal vein,
3. Ligature of the bile duct and clamping and sectioning of the portal vein and hepatic artery,
4. Ligature of the right and left hepatic artery. Clamping and sectioning of the vena cava and portal vein above and below the liver. In this anhepatic phase, venovenous bypass can be used, and the "Gott shunt" cannulae are inserted respectively into the portal vein and femoral vein leading the blood through the "BioiMedicus pump" to the axillary vein (Fig. 4.6).

The clamping of the vena cava and the portal vein causes a great change in the venous return of blood to the heart. Children and most adults support this procedure without a significant drop in cardiac output, but in a proportion of cases, blood flow should be aided by a blood bypass system in which

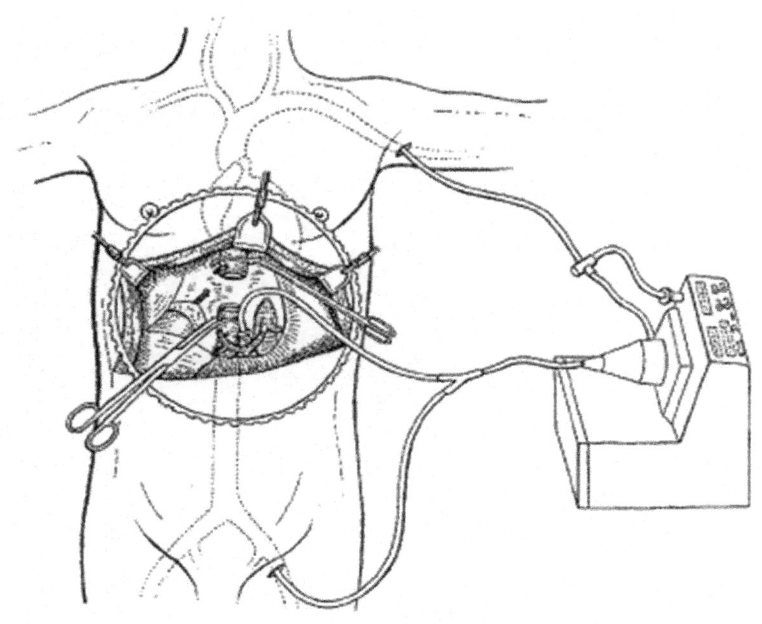

■ **FIGURE 4.6** Recipient surgery: Venovenous bypass system with BioMedicus pump used in the anhepatic phase of liver transplantation.

blood from the inferior vena cava and portal vein is pumped through a pump to the right side of the heart.

Removal of the liver may be associated with considerable bleeding. The ability to maintain the patient in stable conditions despite rapid blood loss is fundamental and anesthesiologists' responsibility. The development of techniques for self-transfusion and rapid infusion of blood has contributed greatly to the current improvement in surgical outcomes. In this sense, patients transplanted by liver tumors should present less complication than those with parenchymal liver disease (cirrhosis).

Alignment of the new orthotopic liver with anastomosis of the terminus-terminus suprahepatic inferior vena cava with continuous polypropylene 3-0. The inferior vena cava and the portal vein are anastomosed, respectively, with prolene 4-0 and 5-0 (Fig. 4.9).

Before the last stitch of the infrahepatic inferior vena cava and portal vein (Haemaccel 500 mL) in the liver through the portal vein to withdraw, the preservation fluid will exit through the infrahepatic inferior vena cava. The venous anastomoses are then completed and the vascular clamps removed to allow perfusion of the liver with portal blood.

■ **FIGURE 4.7** Recipient surgery: Technique of anastomosis of the hepatic artery of the recipient to the celiac trunk of the donor.

■ **FIGURE 4.8** Recipient surgery: Technique of the anastomosis of the accessory hepatic artery of the superior mesenteric to the hepatic artery of the donor.

Suture of the terminal-terminus hepatic artery with prolene 6-0 wire generally at the junction of the hepatic artery with the gastroduodenal artery should be done. If there is arterial variation, suture the superior mesenteric artery with the right hepatic artery in the splenic artery stump of the donor

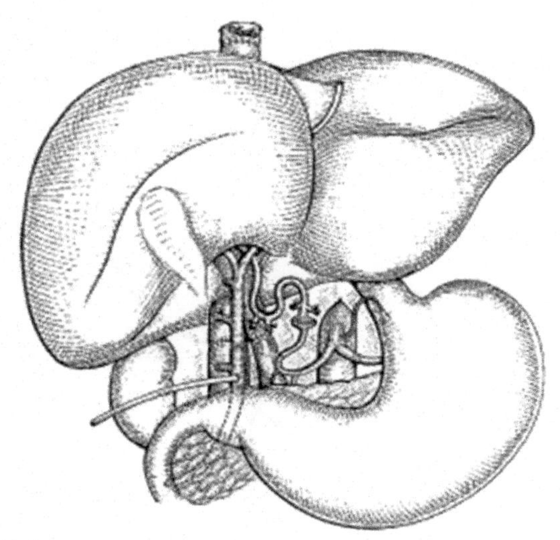

■ **FIGURE 4.9** Recipient surgery: Liver transplanted with the anastomosis of the inferior vena cava, portal vein, hepatic artery, and common bile duct with Kehr drain.

or interpose the superior mesenteric artery between the donor celiac trunk and the hepatic artery of the recipient before the hepatectomy of the recipient (Figs. 4.7 and 4.8).

Reconstruction of the bile duct is done with terminal-terminus anastomosis of the donor duct with that of the recipient with or without a Kehr drain. In cases where the recipient bile duct is sick or absent particularly in cases of sclerosing cholangitis and biliary atresia, the donor bile duct is anastomosed to small bowel in a Roux-en-Y fashion with absorbable vicryl 4-0 (Figs. 4.9 and 4.10).

Wall closing

Check for bleeding and place tubular abdominal drains in the right and left hypochondria. The incision was closed and the patient transferred to the ICU.

In summary, the anastomoses are performed in the following order:

1. Suprahepatic inferior vena cava
2. Infrahepatic inferior vena cava
3. Portal vein
4. Hepatic artery
5. Bile duct

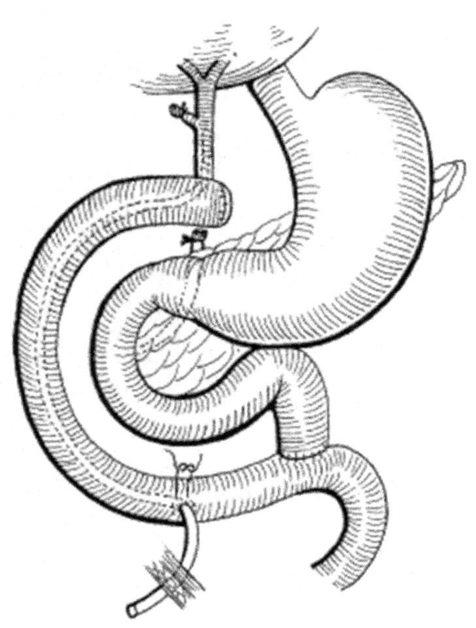

■ FIGURE 4.10 Recipient surgery: Detail of hepatic-jejunal anastomosis with external stent by jejunal loop that can be used for orthotopic liver transplantation.

Principal intraoperative technical care in recipient surgery

1. Portosystemic shunts through the retroperitoneal falciform ligament as well as the peritoneal ligaments may bleed profusely.
2. The hemodynamic changes resulting from the clamping of the portal vein and the inferior vena cava may be life threatening. For this reason, a test is performed at the beginning of the surgery with hepatic vascular clamping (see below). If this is poorly tolerated, the patient is placed in an active venovenous bypass (jugular portocava shunt) with BioMedicus pump (Figs. 4.9 and 4.10).
3. Cardiac arrest may occur after hepatic reperfusion due to cold perfusion fluid, metabolic acidosis, and sudden hypopotassemia that occurs in systemic circulation.

Use of antibiotics in transplantation

Surgical prophylaxis

Routine: Imipenem 7.5 mg/kg (usually 500 mg for adults) as premedication when the anastomoses are completed and every 6 h up to 48 h of post-operatory (P.O).

Patients allergic to penicillin and carriers of resistant staphylococci should receive Ciprofloxacin 200 mg, vancomycin 500 mg, metronidazole 500 mg (as a premedication and every 12 h up to 48 h of P.O) also in patients with hepatic infarction, acute hepatitis, or retransplantation, but continue ciprofloxacin for a further 1 week of P.O.

Prophylaxis of postoperative infection.

Long-spectrum antibiotic for 48 h.

Trimexazole 1 tablet/day on alternate days.

Amphotericin B 1 tablet/day.

Nystatin suspension 5 mL 4×/day via SNG or oral.

Aciclovir 200 mg/day + pyridoxine in patients at risk of tuberculosis (asiatics and alcoholics).

Hemodynamic monitoring and main complications during liver transplantation

Hemodynamic instability is defined as systemic hypotension with or without bradycardia and is frequently found during liver transplantation especially in graft reperfusion. (Figs. 4.1 and 4.5)

There are three basic phases in liver transplantation: hepatectomy (dissection phase), anhepatic phase, and reperfusion phase.

Dissection phase

The most important cardiovascular complications associated with this phase are those related to intravascular volume contraction (blood loss and drainage of large amounts of ascites).

During this phase, it is important to maintain intravascular volume by infusion of blood, frozen fresh plasma, and saline solution depending on the hematocrit and blood coagulation. Sodium chloride should be added in appropriate doses to avoid hypotension (approximately 5–10 mL per liter of infused blood products, depending on hepatic impairment). The objective was to maintain mean blood pressure at MAP >80 mmHg or blood pressure at SBP >110 mmHg at the above normal levels.

If hypotension persists despite elevated filler pressure and normal calcium levels, the use of vasoactive and inotropic drugs may be necessary by adjusting the dose according to cardiac output and peripheral vascular resistance.

At the end of this phase, the surgeon clamps the hepatic pedicles (hepatic hylum and inferior vena cava). If the systolic pressure falls below 70% of its initial value (SBP <80 mmHg or MAP <60 mmHg) in the presence of adequate volume replacement or if ECG changes occur, venovenous bypass is indicated.

Anhepatic phase

This phase begins when the liver of the donor is removed from the box with ice and placed in the abdominal cavity to perform the vascular anastomoses.

The hemodynamic changes of this phase are related to the clamping of the inferior vena cava and to withdrawal of the liver.

The clamping of the inferior vena cava results from a decrease in venous return of 40%–50%, Fig. 4.6 especially in the absence of collateral blood flow. This is accompanied by a large increase in systemic vascular resistance and cardiac output and reduced by half the values of the preanhepatic phase. Despite the decrease in cardiac output, the venovenous bypass and blood pressure were used generally and maintained by a significant increase in systemic vascular resistance Fig. 4.3.

Removal of the diseased liver results in reduced oxygen consumption, which also contributes to decreased cardiac output. Absence of the liver leads to the inability of citrate metabolism: citrate intoxication results in a decrease in ionized serum calcium and in cardiac depression. The decrease of Ca++ to levels of 0.56 mmol/L is associated with depression of cardiovascular function 6.7.8. Hypocalcemia should be corrected before reperfusion.

The body temperature of the patient usually decreases during surgery, especially in the anhepatic phase and during the venous-by-pass, and because the abdominal contents are exposed, there is evaporation of liquid, and the heat is lost in the bypass connections.

Clamping of the portal flow causes increased portal pressure, and this may result in increased intraoperative bleeding.

Attempts to increase blood and fluid filling pressures should be cautious at this stage because of the risk of fluid overload at revascularization when venous return returns to normal.

Postanhepathic phase

Hepatic reperfusion may cause major hemodynamic changes. This phenomenon has been called postreperfusion syndrome 1. Generally, postreperfusion syndrome occurs in 30% of patients who undergo liver transplantation. Although this phenomenon lasts, only a few minutes can be very serious causing cardiac arrest.

The main disturbances are

1. Bradyarrhythmia with a decrease in cardiac index, mean arterial pressure, and systemic vascular resistance.
2. Increase in central venous pressure, pulmonary capillary pressure, and mean pulmonary arterial pressure.
3. The cardiac output increases despite the transient decrease due to falling cardiac index.
4. The new fall in systemic vascular resistance during reperfusion and due to the opening of the new hepatic circulation and the presence of hitherto unknown vasoactive substances that are released into the systemic circulation and coming from the liver of the donor.
5. It is necessary to ensure serum calcium levels before reperfusion and to be at normal levels. After the alterations of the reperfusion, the surgery runs without abnormalities although the continued bleeding can lead to the presence of pathological fibrinolysis.

Specific complications

Metabolic acidosis

Metabolic acidosis that develops during the dissection phase of liver transplantation is usually caused by tissue hypoperfusion of vascular anastomoses present in diseased liver. Correction with sodium bicarbonate is at the discretion of the anesthetist.

After reperfusion, metabolic acidosis generally worsens as a result of rapid release of protons from the ischemic liver and obstruction of splanchnic blood flow. This is usually self-limited and resolves within a few minutes without specific treatment.

Bleeding

Intraoperative bleeding is usually measured by the weight of the gauzes and by the observation of the aspiration bottles, which sometimes underestimate the total bleeding because there are losses that are not measurable. In large bleeds, it is important to transfuse adequate amounts of blood so that blood pressures and central venous pressure are maintained at normal levels. It should be remembered that the packed red blood cells as well as the washed red blood cells must be supplemented with colloids and that the rapid transfusion of whole blood can lead to intoxication by citrate, hypocalcemia, and very rarely hyperpotassemia.

Hypocalcemia

The dosage of ionic calcium should be made at regular intervals during the operation. Levels tend to be low even in the first part of transplantation in patients with impaired hepatic function as a result of the action of citrate on blood derivatives.

Despite the administration of calcium, levels tend to fall into the anhepatic phase when citrate metabolism ceases. If blood loss is not substantial, the levels return to normal after reperfusion without the need for treatment. The ultimate goal is to maintain calcium levels within normal limits throughout the operation mainly in the period preceding liver reperfusion.

Hyperpotassemia

Hyperpotassemia occurs after the opening of the vascular clamps in hepatic reperfusion. It lasts very little time and usually does not require specific treatment. However, we must be prepared to ensure the normal levels of serum calcium. In some cases, when potassium levels are greater than 5.5 mmol/L before reperfusion, it should be corrected with glucose and insulin. After revascularization, the level of serum potassium tends to fall due to uptake by the new liver so supplementation with potassium may be required.

Glycemia

Blood glucose levels tend to rise during transplantation many times because of the glucose administered along with the blood derivatives. Hypoglycaemia may occur at the end of the operation due to increased uptake by the new liver.

Hyperglycaemia generally does not need to be corrected with insulin. Two to five units of insulin can be administered per hour if the blood glucose is above 20 mmol/L.

Arrhythmias in reperfusion

Bradycardia is a common finding in hepatic reperfusion probably caused by hyperpotassemia and the rapid change in the temperature of the sinoatrial knock with the passage of cold hepatic fluid from the newly implanted organ. Irritability and atrial fibrillation are not uncommon usually transient and do not require treatment.

Postreperfusion syndrome refers to the condition of some patients who have a major cardiovascular collapse in hepatic reperfusion with blood pressure remaining at less than 60% of their initial levels for up to 1 minute within 5 minutes of reperfusion. Some studies suggest that prostaglandins, neurotensin, and vasoactive intestinal peptide do not play a major role in this syndrome (Figs. 4.2 and 4.5). Its cause is unknown because its etiology must be multifactorial.

This syndrome is treated with small doses of inotropic vasoconstrictors. However, sometimes the syndrome is so profound that it can lead to cardiac arrest.

The dynamics of waiting list

5.1 LIVER TRANSPLANTATION: WAITING LIST DYNAMICS IN THE STATE OF SÃO PAULO, BRAZIL

São Paulo is the first Brazilian state to perform liver transplantation in 1968 (Machado, 1972). Since then, the recipient waiting list has increased; now approximately 150 new cases per month are referred to the single list at the central organ procurement organization. Official data have shown 37.3 monthly deaths on the waiting list in the state of São Paulo. The number of liver transplants has increased after the creation of São Paulo transplant notification centers but is insufficient to deal with the increasing waiting list. The aim of this study was to demonstrate the performance of our state liver transplantation program and analyze when the number of liver transplantations will meet our waiting list demand.

Materials and methods

We collected official data from State Center of Transplantation—State Secretariat of São Paulo about our liver transplantation program between July 1997 and October 2004. Only cadaveric liver transplantations were recorded; living-related liver transplantation cases were excluded.

The data related to the actual number of liver transplantation (Tr), the incidence of new patients on the list (I), and the number of patients who died in the waiting list (D) in the state of São Paulo since 1997 are shown in Table 5.1.

We took the (Tr) data from Table 5.1 fitting a continuous curve by the method of maximum likelihood, to project the number of transplantations (Tr), in the future. The resultant equation is

$$Tr = 107.07 \ln(\text{year}) + 72.943 \tag{5.1}$$

Mathematical Approaches to Liver Transplantation. https://doi.org/10.1016/B978-0-12-817436-4.00005-9

Table 5.1 Actual number of liver transplantations.

Year	Tr	I	D
1997	63		
1998	160	553	321
1999	188	923	414
2000	238	1074	548
2001	244	1248	604
2002	242	1486	725
2003	289	1564	723
2004	295	1500	671

Tr, the incidence of new patients in the list; I, and the number of patients who died in the waiting list; and D, in the state of São Paulo since 1997.

Results

The results are visualized in Fig. 5.1, in which the number of transplantations performed is fitted to the function above that will be used to project the list size. Note that the number of transplantations since 1997 increased in a nonlinear way, with a clear trend to flattening up to an equilibrium of about 350 cases per year.

We next projected the size of the waiting list (L) by taking into account the incidence of new patients per year (I_t), the number of transplantations carried out in that year (Tr_t), and the number of patients who died in the waiting list (D_t). The dynamics of the waiting list is given by the difference equation:

$$L_{t+1} = L_t + I_t - D_t - Tr_t \tag{5.2}$$

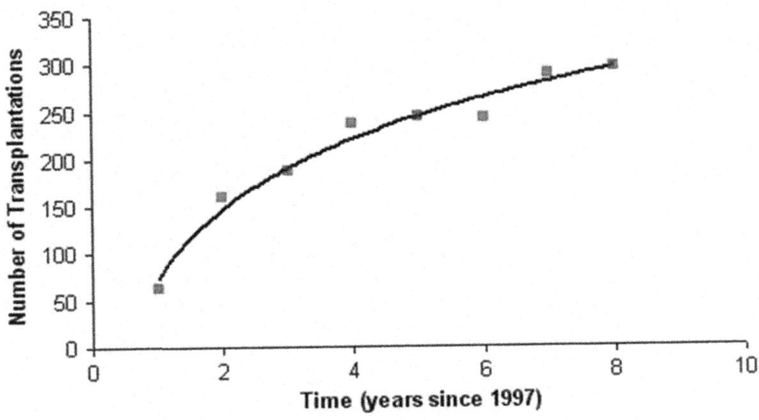

■ **FIGURE 5.1** Fitting curve by the method of maximum likelihood, to project the number of transplantations, Tr, in future time ($P < 0.00000001$).

where the list size at time $t+1$ is equal to the size of the list at the time t, plus the new patients getting into the list at time t, minus those patients who died in the waiting list at time t, and minus those patients who received a graft at time t. The variables I and D from 2004 onward were projected by fitting an equation by maximum likelihood in the same way that we did for Tr.

The waiting list, compared with the number of transplantations, can be seen in Fig. 5.2. Note that, provided present-day conditions, the two projected curves will never meet each other. In other words, the list size grows at a rate much greater than the number of transplantations actually performed.

Conclusions

The biggest challenge facing the field of liver transplantation is the critical shortage of donor organs, which has led to a dramatic increase in the number of patients on the waiting list as well as in their waiting time. Although there is some variability by blood group, most transplant recipients wait well more than 2 years before finally receiving an organ.

To prevent patients from developing life-threatening complications, transplant physicians must therefore make timely decisions to enlist their patients for transplantation. Ideally, liver transplantation should be performed early enough, so that the patient is able to tolerate the surgery, yet sufficiently late in the course of the disease, so that prolonged survival is unlikely without a liver transplant. In practice, however, determining such an optimal time for transplantation may not be that straightforward.

The number of liver transplantations increased 1.84-fold (from 160 to 295) from 1998 to 2004, but the number of patients on the liver waiting list

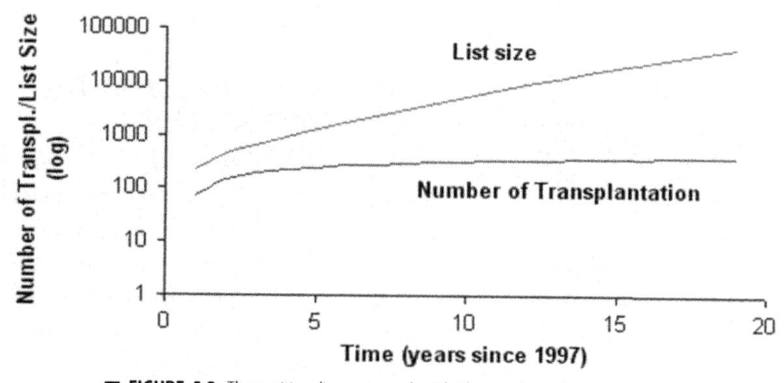

■ **FIGURE 5.2** The waiting list compared with the number of transplantations.

increased 2.71-fold (from 553 to 1500), as would be expected, the number of deaths of listed patients increased 2.09-fold (from 321 to 671; Fig. 5.1). The obvious conclusion is that less than two liver transplantations were done per month per center.

The deaths of listed patients are unacceptable. Many reasons could be pointed to explain these data: (1) The current supply of donor livers is insufficient to meet this need, and organ donation has been stagnant or increased by only a few percent in recent years. (2) There are many liver transplant centers that are doing much fewer transplants than one would expect. (3) The liver transplant teams in the public university hospital are doing much fewer liver transplants than would be expected because of both the full-team capacity of working and the shortage of specialized transplant teams.

Since livers are a limited national resource, our proposal is (1) improve liver donation campaign as we know that we do not already reach the full capacity of donation (currently 7.09 per million inhabitants). (2) Eliminate centers that in the last year performed fewer liver transplantations than would be expected because they have no impact on the waiting list demand. (3) Concentrate funding resources in public university hospitals both to improve the liver transplantation performance and to add more liver transplant teams.

Finally, a word of caution: All of our results are based on a projection of the number of transplantations actually done, assuming that the current trend will persist in the future. Therefore, the conclusions must take this assumption into account.

In conclusion, unless we change the current trend of the number of liver transplantations done in our state, we will see that the projections of Fig. 5.2 will prove correct, that is, we will never meet our waiting list demands in the years to come.

5.2 COMPARING THE DYNAMICS OF KIDNEY AND LIVER TRANSPLANTATION WAITING LIST IN THE STATE OF SÃO PAULO, BRAZIL

As mentioned in the previous chapter, São Paulo is a pioneer Brazilian state on transplantation surgery (Machado, 1972). Kidney and liver transplantation was first performed at Sao Paulo Medical School (Campos Freire et al., 1968). Since then, the patient waiting list for both kidney and liver transplantation has increased now, and approximately, 600 and 150 new cases, respectively, per month are referred to a single list at the central organ procurement organization.

Kidney and liver transplantation has been saving and improving lives for many years. Data have been presented indicating that survival with kidney transplant exceeds survival on dialysis. Because patients survive longer with a transplant than on dialysis, kidney transplantation waiting list is now under great demand in our state.

The gap between the number of transplantable organs from deceased donors and the number of patients awaiting transplantation continues to increase each year. During the past decade, the number of kidney transplants performed has increased, but the number of people developing end-stage renal disease (ESRD) has increased at a greater rate especially; consequently, cadaveric and live donation are not meeting the current demand for organs transplantation.

The number of people waiting for kidney transplantation in our state is approximately 3.04 times the number who receives transplants, considering live and cadaveric kidneys transplants performed. The aim of this study was to compare the performance of our state kidney and liver transplantation program and analyze when the number of transplantations for both will meet our waiting list demand.

Methods and results

We collected official data from State Center of Transplantation—State Secretariat of Sao Paulo about our kidney and liver transplantation program between July 1997 and October 2004. Only cadaveric liver transplantations were recorded, but for kidney transplantations, both cadaveric and living related donors were collected.

The data related to actual number of liver and kidney transplantation (Tr), the incidence of new patients on the list (I), and the number of patients who died in the waiting list (D) in the state of Sao Paulo since 1997 are shown in Table 5.2.

As described in Chaib and Massad (2005), we projected the size of the waiting list, L, by taking into account the incidence of new patients per year, I, the number of transplantations carried out in that year, Tr, and the number of patients who died in the waiting list, D. The dynamics of the waiting list is given by Eq. (5.1), $L_{t+1} = L_t + I_t - D_t - Tr_t$.

Table 5.2 Actual number of liver and renal transplantation.

		Liver			Kidney	
Year	**Tr**	**I**	**D**	**Tr**	**I**	**D**
1997	63			450		
1998	160	553	321	541	1598	492
1999	188	923	414	787	2980	834
2000	238	1074	548	907	3430	1014
2001	244	1248	604	921	3440	1248
2002	242	1486	725	874	1910	976
2003	289	1564	723	891	2544	1080
2004	295	1500	671	1001	2146	695

Tr, the incidence of new patients in the list, I, and the number of patients who died in the waiting list, D, in the state of São Paulo since 1997.

That is, the list size at time $t+1$ is equal to the size of the list at the time t, plus the new patients getting into the list at time t, minus those patients who died in the waiting list at time t, and minus those patients who received a graft at time t. The variables I, and D, from 2004 onward were projected by fitting an equation by maximum likelihood, in the same way that we did for Tr.

The waiting list, compared with the number of transplantations, can be seen in Fig. 5.3. Note that, provided the conditions of the present day, the projected curves will never meet each other. In other words, the list size grows at a rate much higher than the number of transplantations actually done.

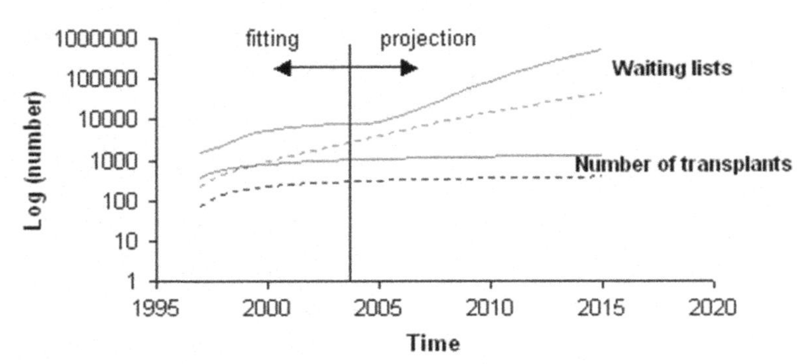

FIGURE 5.3 Comparison between the number of patients in the LTWL and those transplanted.

Conclusions

The biggest challenge facing the fields of transplantation is the critical shortage of donor organs, which has led to a dramatic increase in the number of patients on the waiting list as well as in their waiting time. Kidney and liver transplantation is the treatment of choice for most of both patient with ESRD and chronic or acute hepatic insufficiency (HI). Optimal outcomes occur, respectively, when kidney and liver transplantation is performed as early as possible after the onset of ESRD and HI and risks of both death and graft failure increase with the length of time on dialysis and hepatic clinical supporting measures.

To prevent patients from developing life-threatening complications, transplant physicians must therefore make timely decisions to enlist their patients for transplantation. Ideally, kidney and liver transplantation should be performed early enough so that the patient is able to tolerate the surgery yet sufficiently late in the course of the disease so that prolonged survival is unlikely without kidney and liver transplant. In practice, however, determining such an optimal time for transplantation may not be that straightforward.

The number of kidney and liver transplantations increased approximately 1.85- and 1.84-fold (541−1001 and from 160 to 295) from 1988 to 2004, respectively. On the other hand, the number of patients on the kidney and liver waiting list increased to approximately 2.74 and 2.71(from 1598 to 2741 and 553 to 1500), respectively; as one would be expect, the number of deaths of listed patients increased to approximately 1.97- and 2.09-fold (from 492 to 974 and from 321 to 671), respectively (Fig. 5.3).

The deaths of listed patients are unacceptable. Many reasons could be pointed to explain these data: (1) the current supply of donor livers is insufficient to meet the need, and organ donation has been stagnant or increased by only a few percent in recent years; (2) there are many kidney and liver centers that are doing much fewer transplants than one would expect; and (3) the kidney and liver transplant teams in the public university hospital are doing much fewer kidney and liver transplants than would be expected because of both the full-team capacity of working and the shortage of specialized transplant teams.

When we analyzed only 2003 Brazilian kidney and liver waiting list transplantation, we could see that 4.7 liver transplantations were performed per million inhabitants; we should be performing at least 20 liver transplantations per million inhabitants. This chart was even worse when we looked at the kidney transplantation program. Only 18.5 kidney transplantations were performed per million of inhabitants; we should be performing at least 60 kidney transplantations per million of inhabitants. Based on these data, the necessity of liver and kidney transplantation according with the waiting list would be 33 and 179.2 transplantations/million of inhabitants, respectively (Garcia, 2005).

Because kidneys and livers are a limited national resource, our proposal is as follows: (1) improve organ donation campaign because we know that we do not already reach the full capacity of donation (currently 7.09 per million inhabitants) (Garcia, 2005). (2) Concentrate funding resources in public university hospitals both to improve the kidney and liver transplantation performance and also add more transplants teams. (3) Change the law and start using non–heart-beating donors (NHBD).

NHBD programs remain unpopular despite the potential to increase the donor pool by up to 30% (Varty et al., 1994). A number of legal, ethical, and logistic reasons as well as medical concerns are responsible for this and have even compromised existing NHBD programs (Laskowski et al., 1999).

In recent years, there has been a reevaluation of the use of NHBDs for renal transplantation. Although some studies have shown poorer graft survival for NHBD kidneys (Yokoyama et al., 1994; Feldman et al., 1996), others have demonstrated favorable graft survival compared with heart-beating donors (Kofmann et al., 1993; Wijnen et al., 1995; Nicholson et al., 1997; Gonzalez-Segura et al., 1998) despite the detrimental effects of warm ischemic damage in NHBDs with consequent high rates of delayed graft function.

The use of NHBD offers a large potential of resources for renal transplantation. The process of graft selection involves a significant number of potential grafts being discarded because they are judged to be nonviable. The reported discard rate of kidneys from NHBD is significant, with estimates ranging from 50% to 65% with uncontrolled donors (Metcalfe and Nicholson, 2000).

As far as the NHBD for liver transplantation is concerned, Abt et al. (2003) have emphasized the importance of short cold ischemic time for liver grafts. When the cold ischemic time was less than 8 h, there was 10.8% of graft failure within 60 days of transplantation, which increased to 30.4% and 58.3% when the cold ischemia time was greater than 8 and 12 h, respectively. The combination of warm and cold ischemia in NHBDs appears to make these grafts more susceptible to biliary complications. A greater incidence of ischemic cholangiopathy has been reported (Abt et al., 2003;

Garcia-Valdecasas et al., 1999). D'Alessandro et al. (2000) also recognized increased incidence of ischemic biliary complications in NHBD liver graft recipients from donors older than 40 years of age.

Finally, a word of caution: all of our results are based on a projection of a number of transplantation actually done, assuming that current trend will persist in the future; therefore, the conclusions must take this assumption into account. In conclusion, unless we change the current trend of the number of kidney and liver transplantations performed in our state, we will see that projections of Fig. 5.3 will prove correct, that is, we will never meet our waiting list demands in the years to come.

Chapter 6

Improving grafts allocation

6.1 THE POTENTIAL IMPACT OF USING DONATIONS AFTER CARDIAC DEATH ON THE LIVER TRANSPLANTATION PROGRAM AND WAITING LIST IN THE STATE OF SÃO PAULO, BRAZIL

Introduction

Since liver transplantation was first performed in Brazil at the University of Sao Paulo School of Medicine, the patient waiting list for liver transplantation has increased at a rate of 150 new cases per month. Liver transplantation rose 1.84-fold (from 160 to 295) from 1988 to 2004. However, the number of patients on the liver waiting list jumped 2.71-fold (from 553 to 1500). Consequently the number of deaths on the liver waiting list moved to a higher level, from 321 to 671, increasing 2.09-fold (Chaib and Massad, 2005).

The gap between the number of available organs from deceased donors and the number of patients on the waiting list continues to widen each year all over the word (Austin et al., 2007; Wojcicki et al., 2007; Gruttadauria et al., 2008). This is happening because the number of heart-beating donors is declining, and this is likely to continue for two major reasons: fewer young people are dying as a result of severe injury or catastrophic cerebrovascular events and improvements in the diagnosis and management of severe brain injuries mean that fewer fulfill the brainstem testing criteria.

The biggest challenge facing the field of transplantation is the critical shortage of donor organs, which has led to a dramatic increase in the number of patients on the waiting list and in their waiting time for transplantation (Ahmad et al., 2007; Neuberger et al., 2008).

Mathematical Approaches to Liver Transplantation. https://doi.org/10.1016/B978-0-12-817436-4.00006-0

In a previous work (Chaib and Massad, 2008a,b,c), we projected the size of the waiting list of Sao Paulo State versus the number of transplants carried out in the same period. We demonstrated that the list size grows at a rate much higher than the number of transplants actually performed.

Our latest study found that 4.7 liver transplants were performed per million inhabitants in 2003, but we should have performed at least 33 transplants per million inhabitants.

Cadaver organs from donation after cardiac death (DCD) have been used for decades. Since the introduction of brainstem death criteria in 1968, DCD has been largely abandoned in favor of brain-dead donors (Beecher, 1968). Our country does not use DCD for liver grafts; it uses only heart-beating donor and living-related liver transplantation (Maksoud et al., 1991; Neto et al., 2007).

The aim of this chapter is to analyze, through a mathematical model, the potential impact of using DCD on both our liver transplantation program and waiting list.

Abbreviations

Here, α_i refers to the disease-induced mortality rates; β_i, potentially infective contact rates; γ_i, recovery rates from infection; δ_1, rate of evolution from acute hepatitis to chronic hepatitis; δ_2, rate of evolution to hepatic failure (Model for End-Stage Liver Disease); ε_i; rate of evolution between the Model for End-Stage Liver Disease levels; θ, rate of evolution to hepatocellular carcinoma; Λ, birth rate; μ, natural mortality rate; ξ, rate of transfer to the high-risk group; σ_i, rates of evolution to hepatitis; τ_i, transplantation rates; ϕ_i, rates of getting onto the waiting lists; ω, rate of evolution to other hepatopathies; $Ac(t)$, individuals with acute hepatitis; $Chr(t)$, individuals with intermediate fibrosis stages; *DCD*, donation after cardiac death; *HCC*, hepatocellular carcinoma; $HCC(t)$, individuals with hepatocellular carcinoma; *HCV*, hepatitis C virus; $HCV(t)$, individuals infected with hepatitis C virus who are asymptomatic but already infectious; *MELD*, Model for End-Stage Liver Disease; $MELD_i$, individuals with Model for End-Stage Liver Disease scores; $S(t)$, susceptible individuals; $S_1(t)$, susceptible individuals in the general population; $S_2(t)$, susceptible individuals in a group with a higher risk of acquiring the infection by hepatitis C virus; u, attenuation of β_i affecting people in the nonrisk group; $WL(t)$, individuals on the waiting list for liver transplantation; and $WLTx(t)$, individuals on a secondary waiting list for liver transplantation.

The model

The model was originally proposed to study hepatitis C (Massad et al., 2008a,b) without treatment or vaccination, and it assumes a population divided into 14 states. Susceptible individuals [$S(t)$] are subdivided into two classes, one representing the general population [$S_1(t)$] and the other representing a group with a higher risk of acquiring the infection by hepatitis C virus [$S_2(t)$], such as injecting drug users, recipients of blood and blood product transfusions, and people occupationally exposed to blood and blood products. $S_1(t)$ individuals are transferred to the high-risk group $S_2(t)$ at rate ξ. Both susceptible classes acquire the infection at rates β_i ($i = 1, \ldots, 4$) from four different types of individuals, including infected and asymptomatic but already infectious individuals [$HCV(t)$], individuals with acute hepatitis [$Ac(t)$], and individuals with intermediate fibrosis stages, who are called "chronic state" [$Chr(t)$]. Only fraction u of β_i is effective for people in the nonrisk group. Susceptible individuals can develop other hepatopathies causing liver failure at rate ω; afterward, they are transferred to the state called $Oth(t)$. Infected and asymptomatic individuals [$HCV(t)$] can evolve to acute or chronic hepatitis at rates σ_1 and σ_2, respectively. These individuals evolve to $Chr(t)$ at rate δ_1. Once in the chronic state, individuals can either evolve to hepatocellular carcinoma [$HCC(t)$] at rate θ or evolve to liver failure at rate δ_2. We consider only four levels of the Model for End-Stage Liver Disease (MELD), a scale of hepatopathy that incorporates three widely available laboratory variables, including the international normalized ratio, serum creatinine, and serum bilirubin. Those individuals are denoted $MELD_i$ ($i = 1, \ldots, 4$). Evolution between the MELD levels occurs at rates ε_i ($i = 1, \ldots, 3$). Infected individuals without chronic hepatitis can recover to compartment $R(t)$ at rates γ_i ($i = 1, \ldots, 4$). Individuals from compartments $MELD_i$ ($i = 1, \ldots, 4$) and $HCC(t)$ get onto the waiting list for liver transplantation [$WL(t)$] at rates φ_i ($i = 1, \ldots, 6$). Individuals on the waiting list are eventually transplanted at rate τ_1. We also consider the possibility of graft loss at rate φ_6 and a secondary waiting list [$WLTx(t)$]. From the latter, individuals are eventually transplanted at rate τ_2. Every individual in this population is subjected to mortality rate μ, and additional mortality of those with liver disease occurs at rates α_i ($i = 1, \ldots, 10$), depending on the state. Finally, the susceptible compartment grows at rate Λ, which is composed of the sum of all mortalities to keep the total population constant.

The model's structure is summarized in Fig. 6.1.

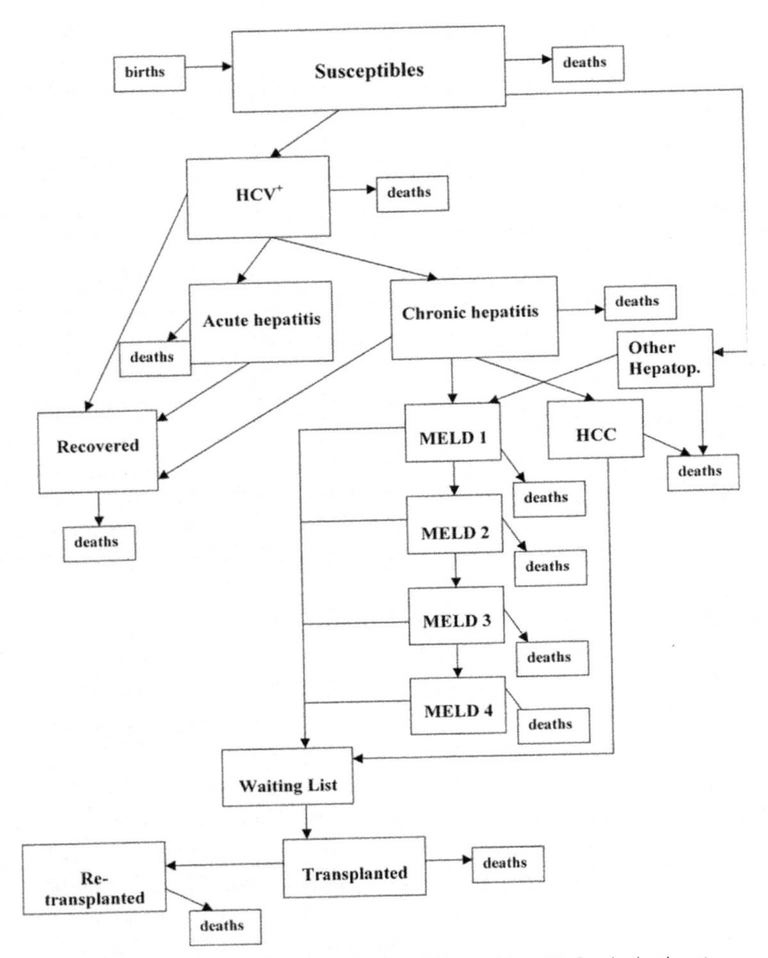

■ **FIGURE 6.1** Model's structure according to the natural history of hepatitis C and other hepatic diseases and organ transplantation. *HCC*, hepatocellular carcinoma; *HCV*, hepatitis C virus; *MELD*, Model for End-Stage Liver Disease.

The model's simulation

The model was simulated with the set of parameters listed in Table 6.1, which were chosen by the Latin hypercube sampling technique (Luz et al., 2003) to reproduce the situation in our state, as described previously (Chaib and Massad, 2008a,b,c).

Table 6.1 Parameters used in the simulations.

Parameter	Biological meaning	Numerical value
Λ	Birth rate	Variable
u	Attenuation of β_i affecting people in the nonrisk group	0.05
ξ	Rate of transfer to the high-risk group	1.0×10^{-5} days^{-1}
μ	Natural mortality rate	3.9×10^{-5} days^{-1}
β_i	Potentially infective contact rates	3.0×10^{-3} days^{-1}
γ_i	Recovery rates from infection	5.0×10^{-5} to 1.0×10^{-4} days^{-1}
σ_i	Rates of evolution to hepatitis	1.0×10^{-3} to 9.0×10^{-3} days^{-1}
δ_1	Rate of evolution from acute hepatitis to chronic hepatitis	7.0×10^{-5} days^{-1}
δ_2	Rate of evolution to hepatic failure (MELD)	5.0×10^{-5} days^{-1}
θ	Rate of evolution to hepatocellular carcinoma	1.0×10^{-6} days^{-1}
φ_i	Rates of getting onto the waiting lists	1.1×10^{-7} to 2.0×10^{-6} days^{-1}
τ_i	Transplantation rates	Variable
ω	Rate of evolution to other hepatopathies	1.3×10^{-6} days^{-1}
α_i	Disease-induced mortality rates	4.2×10^{-5} to 5.9×10^{-5} days^{-1}

In the year 2004 alone, 295 liver transplants were performed in our state, and 671 patients died on the waiting list. When we applied the model, we found out that 265 liver transplants should have been performed, and we expected 674 deaths on the liver transplantation waiting list. These results are in accordance with the reality described previously (Chaib and Massad, 2008a,b,c). On the basis of these values and using the model, we simulated the transplantation rate to predict the increase in the number of liver transplants when cardiac death donors start to be used in our liver transplantation program.

The model's dynamics is described by the following set of equations:

$$
\begin{aligned}
\frac{dS_1(t)}{dt} &= \Lambda - u\beta_1 HCV(t)\frac{S_1(t)}{N} - u\beta_2 Ac(t)\frac{S_1(t)}{N} - u\beta_3 T_A(t)\frac{S_1(t)}{N} \\
&\quad - u\beta_4 Chr(t)\frac{S_1(t)}{N} - (\mu + \omega + \xi)S_1(t) \\
\frac{dS_2(t)}{N} &= \xi S_1(t) - \beta_1 HCV(t)\frac{S_2(t)}{N} - \beta_2 Ac(t)\frac{S_2(t)}{N} - \beta_3 T_A(t)\frac{S_2(t)}{N} \\
&\quad - \beta_4 Chr(t)\frac{S_2(t)}{N} - (\mu + \omega)S_2(t) \\
\frac{dHCV(t)}{dt} &= \beta_1 HCV(t)\frac{(S_1(t) + S_2(t))}{N} + \beta_2 Ac(t)\frac{(S_1(t) + S_2(t))}{N}
\end{aligned}
$$

$$
\begin{aligned}
&+\beta_3 T_A(t)\frac{(S_1(t)+S_2(t))}{N}+\beta_4 Chr(t)\frac{(S_1(t)+S_2(t))}{N}-(\gamma_1+\sigma_1 \\
&+\sigma_2+\mu)HCV(t) \\
\frac{dAc(t)}{dt} &= \sigma_1 HCV(t)-(\delta_1+\gamma_2+\alpha_1+\mu)Ac(t) \\
\frac{dR(t)}{dt} &= \gamma_1 HCV(t)+\gamma_2 Ac(t)+\gamma_4 Chr(t)-\mu R(t) \\
\frac{dChr(t)}{dt} &= \delta_1 Ac(t)+\sigma_2 HCV(t)-(\delta_2+\theta+\alpha_2+\gamma_3)Chr(t) \\
\frac{dMELD_1(t)}{dt} &= \delta_2(Chr(t)+Oth(t))-(\phi_1+\varepsilon_1+\alpha_3+\mu)NT_C M_1(t) \\
\frac{dMELD_2(t)}{dt} &= \varepsilon_1 NT_C M_1(t)-(\phi_2+\varepsilon_2+\alpha_4+\mu)NT_C M_2(t) \\
\frac{dMELD_3(t)}{dt} &= \varepsilon_2 NT_C M_2(t)-(\phi_3+\varepsilon_3+\alpha_5+\mu)NT_C M_3(t) \\
\frac{dMELD_4(t)}{dt} &= \varepsilon_3 NT_C M_3(t)-(\phi_4+\alpha_6+\mu)NT_C M_4(t) \\
\frac{dHCC(t)}{dt} &= \theta Chr(t)-(\phi_5+\alpha_7+\mu)HCC(t) \\
\frac{dWL(t)}{dt} &= \sum_{i,j=1}^{4}\phi_i NT_C M_j(t)+\phi_5 Chr(t)-(\tau_1+\alpha_8+\mu)WL(t) \\
\frac{dTx_1(t)}{dt} &= \tau_1 WL(t)-(\phi_6+\alpha_9+\mu)Tx_1(t) \\
\frac{dWLTx(t)}{dt} &= \phi_6 Tx_1(t)-(\tau_2+\alpha_{10}+\mu)WLTx(t) \\
\frac{dTx_2(t)}{dt} &= \tau_2 WLTx(t)-(\alpha_{11}+\mu)Tx_2(t) \\
\frac{dOth(t)}{dt} &= \omega(S_1(t)+S_2(t))-(\delta_2+\alpha_{12}+\mu)Oth(t) \qquad (6.1)
\end{aligned}
$$

Fig. 6.2 shows the results of the model's simulation. In the figure, the relative reduction of the waiting list is shown as a function of the number of transplants.

Also, in the year 2004, 5100 people died because of accidents and other violent causes in the State of São Paulo (DATASUS, 2008), and 295 of these people were donors of liver grafts that were transplanted. Therefore, 4805 potential liver donors were lost to diverse causes. When we assumed the proportion of potential grafts actually transplanted in the United Kingdom with non—heart-beating techniques (UK Transplant Activity, 2005/2006, 2008) and applied the conservative proportion of 5% to our figures for the State of São Paulo, we ended up with 240 additional liver transplants.

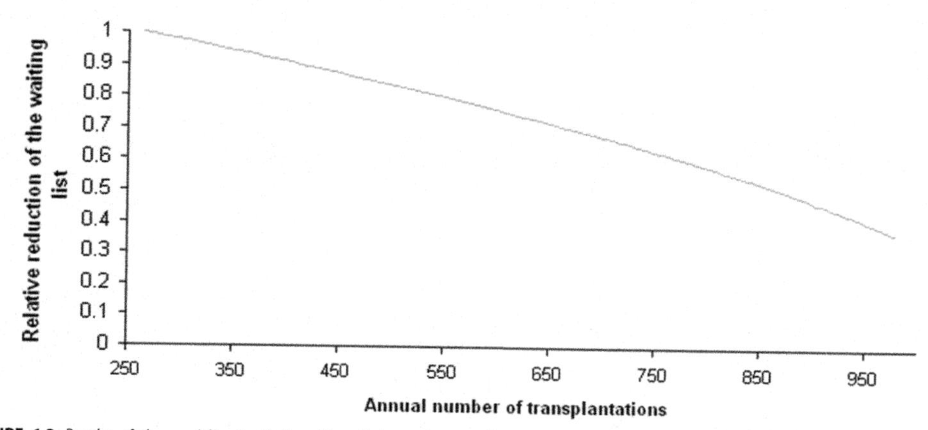

FIGURE 6.2 Results of the model's stimulation. The relative reduction of the waiting list is shown as a function of the number of transplants.

These additional liver transplants, when applied in the model, resulted in a relative reduction of 27% in the size of the waiting list, a proportion in accordance with that obtained by the United Kingdom after the introduction of DCD techniques (UK Transplant Activity, 2005/2006, 2008). This automatically implies a similar 27% reduction of deaths on the waiting list, consequently resulting in 137 lives saved annually according to the model's simulation. In this simulation, we applied a retransplantation rate twice as high as that for heart-beating donors.

If we analyze the projected number of deaths on the liver transplantation waiting list in our state, as already described elsewhere (Chaib and Massad, 2008a,b,c), and compare it with an average reduction of 27% in the annual number of deaths on our liver transplantation waiting list, provided that the number of liver transplants increases by 5% because of a DCD policy, we expect about 41,487 averted deaths on the liver transplantation waiting list over a period of 20 years. The projected number of deaths on the liver transplantation waiting list with or without a DCD policy is shown in Fig. 6.3.

The reduction in the projected number of deaths due to a DCD policy is remarkable.

To estimate the sensitivity of the model to different scenarios for the percent of traumatic deaths that might be converted to DCD donors, we simulate it to mimic the situation in which only 1% of the deaths become DCD donors compared with 10% compared with the 5% presented above. The results are shown in Table 6.2.

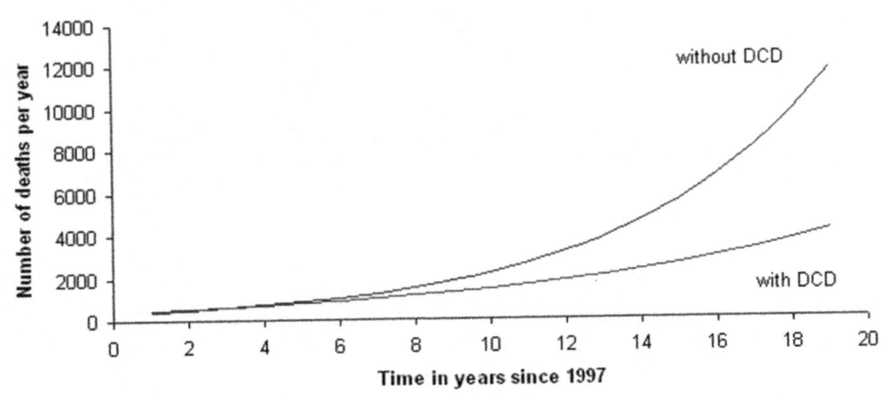

■ **FIGURE 6.3** Projected number of deaths on the liver transplantation waiting list with or without DCD policy. *DCD*, donation after cardiac death.

Table 6.2 Sensitivity analysis.

Percentage of deaths converted to DCD donors	Percentual reduction in the waiting list
1%	8%
5%	27%
10%	37%

Conclusions

São Paulo was the first Brazilian state to perform liver transplantation in Brazil. It was done at Hospital das Clinicas of the University of São Paulo School of Medicine in 1968 (Machado, 1972); however, only in 1985 was the Brazilian liver transplantation program established.

Since then, the recipient liver transplantation waiting list has increased, and approximately 150 new cases per month are currently referred to the single list at the central organ procurement organization. Official data show 37.3 deaths on the liver transplantation waiting list monthly in our state. Although there is some variability by blood group, most of the liver transplant recipients wait far more than 2 years before finally receiving an organ.

Our liver transplantation program itself rose 1.84-fold (from 160 to 295) from 1988 to 2004. However, the number of patients on the liver waiting list jumped 2.71-fold (from 553 to 1500). Consequently the number of deaths on the liver waiting list moved to a higher level, from 321 to 671, increasing 2.09-fold (Chaib and Massad, 2007).

In a previous study (Chaib and Massad, 2007), we found that 4.7 liver transplants were performed per million inhabitants in 2003 when we should have performed at least 33 liver transplants per million inhabitants. Because livers are a limited national resource and we know that we have not already reached the full capacity of donation, we have to improve the liver donation campaign. Basically, we have 7.09 organ donations per million inhabitants, which is far less than one would expect in comparison with other transplant centers worldwide.

In addition, our state liver transplantation policy does not use DCD. Non—heart-beating donor programs remain unpopular, despite the potential to increase the donor pool by up to 30% (Mies et al., 1988). A number of legal, ethical, and logistic reasons and medical concerns are responsible for this and have even compromised existing DCD programs. However, despite these difficulties, DCD is the only hope for reducing the size of our liver transplantation waiting list in the short term.

Because livers are a limited national resource, we propose to (1) improve the organ donation campaign because we know that we have not already reached the full capacity of donation (currently 7.09 per million inhabitants), (2) concentrate funding resources on public university hospitals both to improve liver transplantation performance and to add more transplants teams, and (3) change the law and start using non—heart-beating donors.

When we applied DCD to only 5% of the deaths caused by accidents and other violent causes in our state and compared the result with the projected number of deaths on our liver transplantation waiting list, as already described elsewhere (Chaib and Massad, 2008a,b,c), we found both an average reduction of 27% in the annual number of deaths and 41,487 averted deaths on our liver transplantation waiting list over a period of 20 years.

The projected number of deaths with or without a DCD policy is shown in Fig. 6.3. The reduction in the projected number of deaths due to a non—heart-beating donor policy is remarkable.

Finally, we include a word of caution: all our results are based on a projection of the number of transplants actually performed, assuming that the current trend will persist in the future. Therefore, our conclusions should take this assumption into account.

Unless we change the current trend of the number of liver transplants performed in our state, we will never comply with the demands of our liver transplantation waiting list in the years to come.

In conclusion, the use of DCD in our transplantation program would reduce the pressure on our liver transplantation waiting list, reducing the deaths on it by at least 27%. The projected number of averted deaths based on this model is about 41,487 in the next 20 years, therefore justifying an immediate approach to starting a DCD program in our state.

Donors after circulatory death

7.1 NON—HEART-BEATING DONORS IN ENGLAND

Introduction

The imbalance between supply of organs for transplantation and demand for them is widening all over the world. In the United Kingdom (UK), the figures are not different; by the end of December 2004, there were over 6000 people on the active waiting list for organ transplantation. Recent data suggest that over 400 of these people will die each year before a new organ becomes available (UK transplant, 2004).

When transplantation started, all organs were retrieved from patients immediately after cardiorespiratory arrest, i.e., from non—heart-beating donors (NHBDs). After the recognition that death resulted from irreversible damage to the brainstem by Harvard Medical Committee (CHMSEDBD, 1968) and the subsequent introduction in 1976 (Royal Colleges, 1996) of direct brainstem testing to determine when death has occurred, organ retrieval rapidly switched to patients certified dead after brainstem testing. These heart-beating donors (HBDs) have become the principal source of organs for transplantation for the last 30 years.

The number of HBDs is declining, and this is likely to continue for two major reasons: fewer young people are dying as a result of severe injury or catastrophic cerebrovascular events (WHO, 2010) and improvements in diagnosis and management of severe brain injuries mean that fewer fulfill the brainstem testing criteria.

Although numbers of NHBDs are slowly increasing, they still accounted for only 85 of 750 (11.3%) UK cadaveric donors in 2004—05 (Brook et al., 2004).

The fundamental problem with NHBDs is warm ischemia which may lead to suboptimal transplanted organ function. Developments in organ protection will only lead to more successful outcomes from NHBD if strategies can be devised to keep warm ischemia times (WITs) as short as possible.

Mathematical Approaches to Liver Transplantation. **https://doi.org/10.1016/B978-0-12-817436-4.00007-2**

The aim of this study is to examine clinical outcomes of NHBDs in the past 10 years in the UK as a way of decreasing pressure in the huge waiting list for organs transplantation.

A literature review was performed based on a Medline (Pubmed from 1997 to 2006) search to identify articles on clinical NHBDs in the UK.

Information on the rates of primary nonfunction (PNF), delayed graft function (DGF), acute rejection, and graft and patient survival were registered. Also NHBD technique, perfusion and recipient immunosuppression, were mentioned.

All centers have developed programmes based on the Maastricht protocol (Daemen et al., 1994), which includes the following principles:

1. approval by local medical ethics committee,
2. diagnosis of death by doctors who are independent of the transplant team,
3. the 10-min rule (after declaration of cardiac death, the body is left untouched for a period of 10 min prior intervention), and
4. rapid in situ cooling using a catheter inserted into the aorta, and organ retrieval using standard surgical techniques. Declaration of cardiac death implies irreversible cessation of heart function and the 10-min hands-off period ensures the process of irreversible brain death has begun.

We also recorded the criteria for exclusion of NHBDs, the Maastricht classification, the current criteria for both brainstem death for HBDs and cardiac death for NHBDs, and finally ethics and legal issues involved in the NHBD transplantation.

The first international workshop in Maastricht (Kootstra G, Daemen JHC, Oomen, 1995) defined four categories of NHBD

Uncontrolled

Category I includes victims of accident and suicide (some centers exclude suicide victims from their programs) who are found dead at the scene and resuscitation is deemed pointless (e.g., fatal cervical spine fracture). These are the worst group of potential donors because of the unknown primary WIT.

Category II donors are mostly victims of sudden cardiac (the majority) or cerebral catastrophe who are brought to emergency departments while being resuscitated by ambulance personnel or who died in the department.

Other sources include patients suffering isolated brain injury, anoxia and stroke, and victims of major trauma who died soon after hospital admission.

Controlled

Category III encompasses patients who are dying, often on an intensive care unit. These are the patients awaiting cardiac arrest where the treating clinicians have decided to withdraw treatment and not commence resuscitation for various reasons.

Category IV comprises patients who suffer unexpected cardiac arrest during or after determination of brain death.

Ethics and legal issues

There is an increasing worldwide discrepancy between the availability of and need for organ allografts (Brook et al., 2003a,b). It has been estimated that the number of donor organs could rise by 25% through the expanded use of NHBD (Van der Werf et al., 1988), making it inevitable in an era of universal shortage of donors.

All transplant procurement strategies come with an ethical dimension but this particular process raises multiple concerns, not least of which is ambiguity as to the timing and definition of death.

The principal ethical issues concerning NHBDs programs are the use of in situ perfusion to consent the diagnosis of death by cardiac rather than brainstem criteria and at what time after pronouncement of death the in situ cooling should be started (dead donor rule).

English law does not require consent for prolonged ventilator support or placement of the double-balloon triple-lumen (DBTL) catheter, but catheter placement is an invasive procedure and the family members may not wish for this (Brook et al., 2003a,b).

The dead donor rule is pivotal to NHBD organ donation and states that the donor should not be killed by the act of donation (i.e., the donor must be dead by cardiac criteria at the time of retrieval). This became relevant when a decision is taken to discontinue life support measures in a patient who does not meet the criteria for brainstem death and raises the question of when death should be pronounced after cardiac arrest (Nicholson, 2000).

There is a period when the patient is dead by cardiac criteria but not by brainstem criteria (Iwasaki et al., 1990) and 10 min is considered sufficient for the discrepancy to be corrected. There has been considerable debate over the length of time that the patient should be left (Heineman et al., 1995). For ideal preservation, the kidneys should be perfused as soon as the 10-minute waiting period is over.

These considerations were taken into account at the Maastricht workshop (1995). The conclusions of this discussions was that it was safer to apply the dead donor rule only after a 10-minute period of asystole (Kootstra et al., 1995a,b; Brook et al., 2003a,b; Van der Werf et al., 1988; Nicholson et al., 2000; Iwasaki et al., 1990; Heineman et al., 1995). There can be little doubt that after normothermic ischemia for this period, it would be impossible to restore myocardial function and there will be irreversible loss of all neurological function. Then, the criteria for cardiac and brain death will have been satisfied simultaneously.

Warm ischemia renal injury

The period of warm ischemia is usually defined as the time between cardiac arrest and the start of cardiopulmonary resuscitation.

This could more accurately be described as the absolute warm time because there may be other less obvious periods of warm ischemia. The efficacy of external cardiac massage in achieving renal perfusion is not definitely known and will vary according to how well resuscitation is performed. It is likely that a degree of renal warm ischemia is common during cardiopulmonary resuscitation, and this could be described as relative warm time.

Conversely cardiac massage and ventilation which are effective in oxygenating the kidneys after a period of circulatory standstill may also be deleterious by initiating the reperfusion injury syndrome.

Warm ischemia is known to be a major determinant of renal function after kidney transplantation. However, the amount of reversible warm ischemic injury that the human kidney can sustain is still not known for certainty. Most human NHBD kidney transplant protocols exclude kidneys with such prolonged warm times, the usual cutoff being in the 30–45 min (Daemen et al., 1997a,b,c; Dunlop et al., 1995).

Renal allografts from controlled NHBDs sometimes function immediately suggesting that the human kidney tolerates short periods of warm ischemia quite well (Kosaki et al., 1991). In uncontrolled NHBDs, the warm time is not always accurately known. It may need to be determined by taking a history from relatives, ambulance staff, and medical personnel, and in some cases this will provide only an approximate estimate.

Surgical technique

Leicester model for NHBD (in situ kidney perfusion/cooling and kidney retrieval in NHBD)

NHBD kidneys were perfused and cooled in situ using a DBTL aortic catheter placed via femoral artery cut down in the groin. The technique as described originally by Garcia-Rinaldi et al. (1975) involves insertion of the DBTL catheter into the abdominal aorta via femoral arteriotomy. Inflation of the caudal balloon at the aorta bifurcation and inflation of the cranial balloon isolates the segment of aorta from which the renal arteries originate (vertebral level L1). A plain abdominal radiograph can be taken to show the position of the catheter, which is radioopaque.

Wheatley et al. (1996) modification is to mix a small amount of Conray (May and Baker, Dagenham, UK) radiocontrast dye with saline injected into the balloons to allow easier identification of the balloons on radiography. Using a mixture of Conray and saline in a ratio 1:10 rather than neat Conray makes injection of the viscous liquid much easier.

The balloons were inflated with radiographic contrast medium and the positioning checked by plain abdominal X-ray (Wheatley et al., 1996).

The system was vented by placing a Foley catheter into the inferior vena cava via the right femoral vein. Following in situ renal perfusion, the NHBD was transferred to an operating theater for bilateral donor nephrectomy.

The perfusion fluid used for NHBD was a 10—20 L Marshall's hyperosmolar citrate solution at 4 C infused under gravity and all kidneys were held in static ice storage before transplantation.

Newcastle upon-tyne model for NHBD (NHBD retrieval: pumping perfusion system)

In situ organ perfusion was performed by cannulating the femoral artery using a DBTL cannula (TXF Medical, High Wycombe, UK). The preservative solution was cold (4—8 C) heparinized (1000 IU/L) Marshall's solution. The venous venting was through placement of a cannula into the femoral vein. No radiological confirmation of placement was done. After retrieval, the kidneys were cold stored and transported to the hospital where machine perfusion and viability assessment were carried out (Balupuri et al., 2000a).

Pumping perfusion system

A Belco BL 760 blood pump module was used for perfusion. One pump in the system provided fluid to the renal artery and the other retrieved it through a heat exchanger. The temperature of the system was maintained between 4 and 9 C. The pump was capable of delivering a flow rate of 28–480 mL/min and pressure was maintained at 45–60 mm Hg. Thus, a closed system of perfusion was achieved.

A viable kidney tends to have a flow rate of more than 50 mL perfusate/minute per 100 g of kidney and a glutathione S-transferase level less than 200 IU/L per 100 g kidney (Balupuri et al., 2000b).

Cambridge model for NHBD

Following certification of death on the ward, the donor was transported to the operating theater. Laparotomy was performed, a large bore (22F) aortic cannula was inserted, and the kidneys were perfused in situ with at least 3L of cold University of Wisconsin (UW) solution containing 20,000 units of heparin per liter.

The supraceliac aorta was cross clamped to facilitate perfusion of the kidneys, and the venous system was decompressed via inferior vena cava. To provide additional cooling, 2 L of crushed frozen saline were applied around the kidneys during perfusion. Donor nephrectomy was then performed (Sudhindran et al., 2003).

London model for NHBD

An interval of 15 min after cardiac arrest was always allowed before cold in situ perfusion or crash retrieval was commenced. All donors were cooled using Soltran Kidney Perfusion Solution (Baxter Medical, Houston, TX, USA) with an additive of 20,000 IU of heparin via femoral access using a DBTL inserted into the femoral artery. Cannulas were placed only after consent of relatives was given.

The kidneys of the controlled donors were perfused via rapid aortic cannulation after laparotomy or crash retrieved with no flush and thereafter perfused with ice-cold solution on the bench.

Pulsatile kidney perfusion with the RM3 pump (Water Medical Systems, Rochester, MN, USA) for organ assessment and preservation of up to 12 h was used.

Results

During 10-year period analyzed, NHBDs were used mainly for kidney transplantation in four centers in the UK as summarized in Table 7.1.

NHBDs in the UK centers according to the Maastricht classification are shown in Table 7.2.

On comparing the results of NHBD with HBD relating to PNF, only 3 out of 24 articles have shown significant differences. As far as DGF is concern 8 out of 24 articles have shown significant differences. Acute rejection was reported significant in only 1 out of 24 articles (Table 7.3).

Table 7.1 England non—heart-beating donors' transplant details.

Author	Year	Center	No. of patients	Organ
Nicholson et al. (1997)	1997	Leicester	30	Kidney
Butterworth et al. (1997)	1997	Leicester	37	Kidney
Nicholson et al. (2000)	2000	Leicester	77	Kidney
Balupuri et al. (2000a)	2000	Newcastle upon-Tyne	15	Kidney
Balupuri et al. (2001)	2001	Newcastle upon-Tyne	28	Kidney
Metcalfe et al. (2001)	2001	Leicester	72	Kidney
Gok et al (2002a)	2002	Newcastle upon-Tyne	43	Kidney
Gerstenkorn et al. (2002)	2002	London	202	Kidney
Gok et al. (2002b)	2002	Newcastle upon-Tyne	46	Kidney
Sudhindran et al. (2003)	2003	Cambridge	42	Kidney
Gok et al. (2003)[33]	2003	Newcastle upon-Tyne	25	Kidney
Brook et al. (2003a,b)	2003	Leicester	55	Kidney
	2003	London	41	Kidney
Gok et al (2004a)	2004	Newcastle upon-Tyne	72	Kidney
Brook et al. (2004)	2004	[a]	285	Kidney
Gok et al. (2004b)[38]	2004	Newcastle upon-Tyne	02	Kidney
Wilson et al. (2005)	2005	Newcastle upon-Tyne and Leicester	51	Kidney
Bains et al. (2005)	2005	Leicester	37	Kidney
Navarro et al (2006a)	2006	Newcastle upon-Tyne	05	Kidney
Gok et al. (2006)	2006	Newcastle upon-Tyne	19	Kidney
Navarro et al. (2006b)	2006	Newcastle upon-Tyne	81	Kidney
Sohrabi et al (2006a)	2006	Newcastle upon-Tyne	05	Kidney
Sohrabi et al. (2006b)	2006	Newcastle upon-Tyne	36	Kidney
Muiesan et al. (2006)	2006	London	07	Liver

[a]Leicester, Cambridge, London, Newcastle upon-Tyne (combined results of renal NHBD transplantation in the UK from 1988 to 2001).

Table 7.2 Non–heart-beating donors in England centers according to the Maastricht classification.

Center	NHBD* not mentioned	NHBD uncontrolled	NHBD controlled	NHBD total
	Category (n = 365)	Categories I and II (n = 575)	Categorie III and IV (n = 398)	(n = 1334)
Cambridge	–	0	42	42
Leicester	144	144	20	308
London	202	8	40	250
Newcastle upon-Tyne	19	178	201	398
Cambridge, Leicester, London	–	217	68	285
Newcastle upon-Tyne				
Leicester, Newcastle upon-Tyne	–	24	27	51

*NHBD (non–heart-beating donor)

Table 7.3 Comparison of the results of NHBD/HBD in England relating to primary nonfunction, delayed graft function, and acute rejection.

Author	NHBD/HBD	PNF (%) NHBD/HBD	DGF (%) NHBD/HBD	AR (%) NHBD/HBD
Nicholson et al. (1997)	30/114	3/13(s)	25/87(s)	48/27(s)
Butterworth et al. (1997)	37/91	11/1(s)	100/28(s)	27/36(ns)
Nicholson et al. (2000)	77/224	9.1/2.7(s)	84.3/21(s)	28.6/32.6(ns)
Balupuri et al. (2000a)	15/nm	6.6/-	66.6/-	nm
Balupuri et al. (2001)	28/nm	3.57/-	91.6/-	nm
Metcalfe et al. (2001)	72/192	7/4(ns)	80/19(s)	24/31(ns)
Gok et al (2002a)	43/nm	–	–	–
Gerstenkorn et al. (2002)	202/nm	21.8/-	82.3/-	13.1/-
Gok et al. (2002b)	46/46	8.7/2.2	94.9/41(s)	52.2/41.3(ns)
Sudhindran et al. (2003)	42/84	0/2	50/17(s)	33.3/40.5(ns)
Gok et al. (2003)[33]	25/nm	2/nm	75/nm	-
Brook et al. (2003a,b)	55/69	5/3	93/17	24/23(ns)
	41/0	14.6	80	–
Gok et al (2004a)	72/nm	7.85/-	75.6/-	41.4/-
Brook et al. (2004)	285/0	15	79.7	41
Gok et al. (2004b)[38]	2/0	0/0	100/0	100/0
Wilson et al. (2005)	51/nm	2/nm	7/nm	–
Bains et al. (2005)	37/75	0/0	31/8(s)	7/27(ns)
Navarro et al (2006a)	81/-	nm	nm	nm
Gok et al. (2006)	19/15	nm	57.9/45.5(s)	36.8/20(s)
Navarro et al. (2006b)	5/-	nm	nm	nm

Table 7.3 Comparison of the results of NHBD/HBD in England relating to primary nonfunction, delayed graft function, and acute rejection. *continued*

Author	NHBD/HBD	PNF (%) NHBD/HBD	DGF (%) NHBD/HBD	AR (%) NHBD/HBD
Sohrabi et al (2006a)	5/-	0/-	80/-	nm
Sohrabi et al. (2006b)	36/-	6.3/-	nm	35/-
Muiesan et al. (2006)	07/0	0	0	3

AR, *acute rejection;* DGF, *delayed graft function;* HBD, *heart-beating donor;* NHBD, *non—heart-beating donor;* nm, *not mentioned;* ns, *not statistically significant;* PNF, *primary nonfunction;* s, *significant.*

Graft survival was equal for 1, 2, 3, and 5 years when comparing NHBD with HDB except in one article that has shown significant differences in 3 and 5 years time (Table 7.4).

Patient survival was equal in 1, 3, and 5 years when comparing NHBD with HBD as shown in Table 7.5.

UK centers and variations of NHBD technique, use of DBTL, machine perfusion, type of organ perfusion solution, as well as immunosuppression are demonstrated in Table 7.6.

Table 7.4 Graft survival in NHBD in England.

		1 y(%)	2 y(%)	3 y(%)	5 y(%)
Nicholson et al. (1997)	NHBD	78	75	73	—
Butterworth et al. (1997)	HBD	90	85	82	
Nicholson et al. (2000)	NHBD	73	73	73	—
Balupuri et al. (2000a)	HBD	83[a]	79[a]	77[a]	
Balupuri et al. (2001)	NHBD	85[a]	87[a]	87[a]	79[a]
Metcalfe et al. (2001)	HBD	85	82	80	75
Gok et al (2002a)	NHBD	91.7	—	—	—
Gerstenkorn et al. (2002)	NHBD	88.1	—	—	—
Gok et al. (2002b)	NHBD	81	—	77	73
Sudhindran et al. (2003)	HBD	86[a]		78[a]	65[a]
Gok et al. (2003)[33]	NHBD	89.6	—	—	—
Brook et al. (2003a,b)	HBD	91.4[a]			
	NHBD	86.9	—	75.5	65.5
Gok et al (2004a)	NHBD	—	—	82.9	—
Brook et al. (2004)	NHBD	79	—	75	69
Gok et al. (2004b)[38]	NHBD	89.6	—	89.6	—
Wilson et al. (2005)	HBD	91.4[a]		91.4[a]	
Bains et al. (2005)	NHBD	84	—	80	74

Continued

Table 7.4 Graft survival in NHBD in England. *continued*

		1 y(%)	2 y(%)	3 y(%)	5 y(%)
Navarro et al (2006a)	HBD	89[a]		85[a]	80[a]
Gok et al. (2006)	NHBD	88	–	84	84
Navarro et al. (2006b)	HBD	82[a]		73[b]	62[b]
Sohrabi et al (2006a)	NHBD	92.1	–	–	–
Sohrabi et al. (2006b)	NHBD	–	–	95	–
Muiesan et al. (2006)	NHBD	84	–	–	–

HBD, *heart-beating donor;* NHBD, *non—heart-beating donor;* y, *year.*
[a]*not statistically significant.*
[b]*statistically significant.*

Table 7.5 Patient survival in NHBD in England.

		1 y (%)	3 y (%)	5 y (%)
Gok et al. (2002)	NHBD	87.9	87.9	–
	HBD	89.7	89.7	
Gerstenkorn et al. (2002)	NHBD	93.1	85.3	76.2
Sudhindran et al. (2003)	NHBD	91	91	84
	HBD	94	92	90
Gok et al. (2004)	NHBD	92.1	–	–
Muiesan et al. (2006)	NHBD	87	–	–

HBD, *heart-beating donor;* NHBD, *non—heart-beating donor;* y, *year.*

Table 7.6 England centers and variations of non—heart-beating donor programmes.

Center	NHBD technique	In situ perfusion	Immunosuppression
Cambridge	Laparotomy + aortic cannula	UW	CSA/AZA/PRED
Leicester	Femoral arteriotomy + DBTL catheter	Marshall's	CSA/AZA/PRED or TAC/MMF/PRED
London	Kidney—femoral arteriotomy + DBTL	Marshall's	CSA/AZA/PRED
	Liver—PV cannulation	UW	TAC/PRED
Newcastle upon-Tyne	Femoral arteriotomy + DBTL catheter + MP system	Marshall's	Ab induction; TAC/MMF/PRED

Ab, *interleukin-2 receptor antibody;* AZA, *azathioprine (1 mg/kg/d);* CSA, *cyclosporine (7 mg/kg/d);* DBTL, *double-balloon triple-lumen;* MMF, *mycophenolate mofetil (2g b.i.d.);* MP, *machine perfusion;* PRED, *prednisolone (20 mg/d reducing to 5 mg/d at 6 months);* PV, *portal vein;* SKPS, *Soltran kidney perfusion solution;* TAC, *tacrolimus (0.1 mg/kg/d);* UW, *University of Wisconsin.*

Conclusions

Cadaveric organs from NHBD have been used for decades. Since the introduction of "brainstem death" criteria in 1968, NHBDs have been largely abandoned in favor of brain dead donors (Beecher, 1968) NHBD offers a large potential of resources for renal transplantation. The process of graft selection involves a significant number of potential grafts being discarded because they are judged to be nonviable. The reported discard rate of kidneys from NHBD is significant, with estimates ranging from 50% to 65% with uncontrolled donors (Metcalfe and Nicholson, 2000).

NHBD programs remain unpopular despite the potential to increase the donor pool by up to 30% (Varty et al., 1994). A number of legal, ethical, and logistic reasons as well as medical concerns are responsible for this and have even compromised existing NHBD programs (Laskowski et al., 1999).

Legal and ethical issues such as cannulation of the femoral vessels for in situ cooling prior to consent by the relatives and an undefined interval of no-touch between cardiac arrest, declaration of death, and organ resuscitation/preservation efforts remain an unsolved problem in the UK. If relatives were present, the consent rate for organ donation after cardiac arrest was more than 70%.

In addition, the logistic requirements for a successful, financially efficient NHBD program are immense and require good planning, management, and organization. Transplant surgeons and coordinators must be within close reach of potential donor locations in emergency departments, intensive care units, hospital wards, and hospices. Furthermore, a sufficient number of trained staff members are required to share a rota for uninterrupted on-call cover. This might be possible only for a large transplant units or particularly dedicated centers. Organ preservation by perfusion is an additional issue that requires further expertise and is expensive (Alvarez-Rodrigues et al., 1995), but it might be the only reliable technique for assessing the viability of NHBD kidneys and marginal donor organs (Tesi et al., 1993).

The 125 NHBD transplants performed in 2005–06 (82 kidneys only; 36 kidney/liver; 1 kidney, thoracic, and liver, 2 kidney, liver, and pancreas, and 4 liver only) rose 44% comparing with 87 NHBD in 2004–05 but still represent a small proportion of the total transplant activity on the UK (UK Transplant Activity, 2005/2006).

The average NHBD retrieval rate for 2001 was 1.3 million population (pmp) for the four centers reported here. This compares with an average national rate for HBD kidneys of 23.5 pmp (UK Transplant Activity, 2001). Although the number of NHBD transplants is small, the potential is greater; the most encouraging figures come from Daemen et al. who report 40% of their kidneys were accounted for by NHBDs.

Varty et al. (1994) reported 38% of donors were NHBD in 1991, and later Nicholson (1996) reported that NHBD accounted for 21% of transplant activity. Light et al. (2000) stated that the number of NHBD transplant opportunities equals that of HBD, while other authors have suggested that there are twice as many potential NHBD as there are HBD (Cho et al., 1998). It has been suggested that if all potential NHBD kidneys were retrieved, waiting lists for kidney transplantation could be eliminated (Terasaki et al., 1997).

Two methods of harvesting of NHBDs have been advocated to decrease ischemic insult: (1) procurement methods include in situ intravascular cooling (universal), intraperitoneal lavage, and cooling (used in the United States), hypothermic and normothermic partial cardiopulmonary bypass (used in Spain), and hypothermic extracorporeal membrane oxygenation (used in Taiwan) and (2) machine preservation methods include pulsatile intermittent perfusion and continuous-and-constant perfusion.

Ischemic injury is almost inevitable in NHBD renal grafts and primarily results from prolonged hypotension, cardiac arrest, and warm ischemia during the primary WIT. Development of allograft dysfunction can result in DGF and PNF.

As far as the NHBD for liver transplantation is concerned, Muiesan et al. (2006) based their donor selection on several factors including age less than 40 years, short intensive care unit stay, normal liver function tests, short interval from withdraw of therapy to cardiac arrest, short WIT and good appearance, and perfusion of the liver, as judged by an experienced transplant surgeon. When the majority of these parameters were satisfied, they provided organs suitable for bench segmental reduction.

They started the procedure of reduction/splitting as soon as the recovery team had returned to their center. Care was taken to keep the temperature of the University of Wisconsin (UW) solution at 4 C and to avoid rewarming of the graft. One split procedure was performed. The left lateral segment was transplanted with 7 h of cold ischemia and provided excellent immediate function with an International Normalized Ratio of 1.2 on the second postoperative day. The adult recipient who received the right lobe split was transplanted sequentially with a cold ischemia time of 14.3 h and experienced PNF. He died of multiorgan failure after emergency retransplantation.

Subsequent reduction of NHBD grafts was performed with transplantation only of the left lateral segment or left lobe in children (Muiesan et al., 2006).

The recent largest retrospective study comparing outcomes of adults' recipients of NHBD and HBD hepatic allografts between 1993 and 2001 based on the United Network of Organ Sharing database confirmed their greater vulnerability to cold ischemia and the higher incidence of PNF among a group of 144 NHBD livers (Abt et al., 2004).

Abt et al. (2003) have emphasized the importance of short cold ischemic time for liver grafts. When the cold ischemic time was less than 8 h, there was 10.8% of graft failure within 60 days of transplantation which increased to 30.4% and 58.3% when the cold ischemia time was greater than 8 and 12 h, respectively. The combination of warm and cold ischemia in NHBDs appears to make these grafts more susceptible to biliary complications. A higher incidence of ischemic cholangiopathy has been reported (Garcia-Valdecasas et al., 1999).

D'Alessandro et al. (2000) also recognized increased incidence of ischemic biliary complications in NHBD liver graft recipients from donors older than 40 years of age.

To date, there have been no vascular complications and no evidence of major biliary complications with particular reference to ischemic cholangiopathy in these children after a mean follow-up of 19 months (Muiesan et al., 2006).

Reddy et al. (2004) described various cytoprotective strategies involving administration of drugs before cardiac arrest that have been successfully used in NHBD liver transplantation.

Kidneys from different Maastricht categories recovered at different rates although they were all similar at 3 months. Kidneys from NHBDs have undergone more ischemic insult than kidneys from HBDs. This occurs because of the prolonged early (primary) warm ischemia.

For Maastricht category II donors, in this the period between collapse and effective resuscitation, variably effective resuscitation and the "no-touch" period before perfusion should be performed.

For category III donors, after withdrawal of support there can be an agonal period of hypotension followed by the "no-touch" period and then perfusion.

Category IV donors are known to be brainstem dead and are the closest to HBDs. They often arrest in theater, therefore the primary ischemic time does not include the "no-touch" period. Thus, the primary warm ischemic insult is likely to be maximal for category II and minimal for IV with III in between (II greater than III greater than IV) (Gok et al., 2004a,b).

With maximal primary warm ischemic damage comes the increasing chance of in situ thrombosis, greater oxygen debt, and increasing acidosis and reduction of cellular adenosine triphosphate. Viability testing excludes those organs that are so badly damaged they are not likely to function. As a consequence, the prolonged warm ischemia and reperfusion syndrome is likely to be more severe for kidneys from category II donors as opposed to III or IV.

A contrary argument has been put forward that a person who is hospitalized and in an intensive therapy unit for a protracted period is more likely to have inotropic support and sepsis which may result in unmeasured hypoperfusion of kidneys. Therefore, kidneys from a sudden collapse (category II) are possibly healthier than those from a hospitalized patient (Brook et al., 2003a,b).

Gok et al. (2004) demonstrated that the PNF rate is slightly higher for category II donors (13.5%) than category III (2.2%) and category IV (0%, $p < .05$). In addition, DGF incidence is greatest for category II donors (83.8%), less for category III (67.4%), and best for category IV (0%).

The introduction of pulsatile hypothermic machine perfusion of NHBD kidneys, along with viability testing, has resulted in a viable graft, acceptable survival rates with DGF in recipients. The main advantages of machine perfusion are the provision of metabolic support via replenishment of ATP by high energy yielding phosphate bonds from the perfusate (UW solution). Apart from the mechanical flush provided by continuous perfusion, the natural ischemia–induced capillary vasoconstriction is reversed as demonstrated by the reduction of intrarenal vascular resistance. This in turn is reflected in lower incidence of DGF. Machine perfusion also allows monitoring and assessment of viability in an NHBD kidney graft.

Opelz and Terasaki concluded no added advantage of machine perfusion in HBD (Opelz et al., 1981). Machine perfusion fell into disrepute due to its complex logistics needs (Baxby et al., 1974; Scott et al., 1969). It was also reported in some cases that machine perfusion was damaging to the graft (Newman et al., 1975).

Renewed interest in this method of preservation came about through the use of marginal organs and improved preservative solutions (UW). With the improved solutions, the incidence of DGF could be reduced using kidneys from HBDs or those with prolonged cold ischemia (Burdick et al., 1997; Kwiatkosk et al., 1996).

Increased demand for viable organs has led to a recent upsurge in retrieval from the NHBDs. The difficulty experienced here was the viability of such kidneys. As the primary WITs are prolonged with such donors, viability assessment and organ modulation have been done by machine perfusing the NHBD kidneys before transplantation (Backman et al., 1988; Casavilla et al., 1995; Daemen et al., 1997; Kootstra et al., 1997).

The measurement of intrarenal vascular resistance and alpha GST is practiced by the Maastricht group, who has increased their donor pool by 20% (Kootstra et al., 1991). Most of the studies have demonstrated a beneficial effect of pulsatile perfusion in NHBD kidneys (Matsuno et al., 1998). Machine perfusion has been shown to improve the graft function in cases of marginal kidneys as well as those with prolonged cold ischemic times.

In the setting of NHBD kidney transplantation, PNF represents the transplant of an organ irreparably damaged by warm ischemia. WIT correlates well with graft damage and there is a point at which organs become nonviable. There is no strict maximum WIT in NHBDs beyond which transplantation is contraindicated. Limits of 30 min (Haisch et al., 1997), 35 min, and 45 min (Light et al., 1996) have been advocated. Functional recovery in animal models has been achieved after substantially longer periods of 120 min (Haisch et al., 1997) and 140 min (Matsumo et al., 1994).

Renal transplants with prolonged WIT, that is primary warm ischemia, demonstrated a relative increase in free radical generation during NHBD renal transplantation. Use of traditional tissue injury markers, LDH, AST, and lactate, and the specific markers of tissue injury, Ala-AP and FABP, during kidney transplantation complemented the finding of free radical injury in NHBD renal transplants. Combined markers enabled the monitoring of different types of cell injury. This correlates with the high incidence of acute tubular necrosis and subsequent DGF in NHBD renal transplants.

Reperfusion injury is such that antioxidant strategies may be of benefit. Potential measures that could be used are nitric oxide addition (Thabut et al., 2001) and free radical scavengers (e.g., allopurinol, superoxide dismutase, catalase, and dimethyl sulfoxide) with antioxidant supplementation (e.g., glutathione, vitamin E and gingko biloba) (Biasi et al., 2002; Schauer et al., 2004), cytokine—chemokine suppressors, adhesion molecule blockers

(Selzner et al., 2003; Mallick et al., 2004), and neutrophil–endothelial cell blockade (Ortiz et al., 2003; Demertzis et al., 1999).

The role of free radicals in the pathophysiology of ischemia-reperfusion injury in transplantation is increasingly recognized. Free radical formation appears to occur in two phases, characterized as reperfusion-mediated injury (early and short-lived, that is seconds to minutes) and neutrophil-mediated injury (late and long-lived, that is minutes to an hour). In reperfusion-mediated injury, adenosine triphosphate depletion during hypoxia results in the accumulation of hypoxanthine which is converted to free radicals (superoxide and hydrogen peroxide radicals) upon reperfusion by xanthine oxidase activation (Demertzis et al., 1999; Castaneda et al., 2003). In neutrophil-mediated injury, reperfusion results in the activation of neutrophil NADPH oxidase and the release of further free radicals (superoxide, hydrogen peroxide, and hydroxyl radicals).

In kidney allograft, free radicals can be formed throughout the renal parenchyma in the intracellular, intravascular, and injury compartments. Ischemia-reperfusion injury consist of 2 injury mechanisms namely ischemia characterized as oncotic necrosis and reperfusion injury. In the clinical setting, this damage may delay the recovery of renal function prolong or complicate postoperative recovery or precipitate immunological reactions involved in the subsequent rejection reaction.

In practice, allowable maximum WIT varies in a qualitative manner, i.e., a young and previously fit donor may be allowed a longer WIT than an older donor. The PNF average rate in the current study was 6.17% which is consistent with the 8%–15% reported previously from NHBDs (Daemen et al., 1996; Wijnen et al., 1995). This is higher than 2%–5% quoted for HBD kidneys (Butterworth et al., 1997; Schlumpf et al., 1996). Shiroki et al. (1998) claimed that PNF rate is highest in those NHBD kidneys with warm time of >30 min. Tanabe et al. (1998) disagreed stating that warm time bore no relationship to PNF rate but no WIT was >30 min in their study. Brook et al. (2003a,b) showed no episode of PNF in controlled donors with a 20% PNF rate in uncontrolled NHBD kidneys, indicating an association of warm time with PNF.

The reported incidence of PNF after transplantation using kidney from NHB donors varies from 4% to 40%, although most studies do not make distinction between kidneys from controlled and noncontrolled NHB donors when describing the incidence of PNF (Balupuri et al., 2000a,b; Nicholson et al., 2000; Cho et al., 1998; Wijnen et al., 1995; Tanabe et al., 1998; Weber et al., 2002; Daemen et al., 1997; Olson et al., 1996). The PNF is probably a consequence of ischemic cortical necrosis Brook et al. (2003a,b).

The perceived risk of PNF and poor graft outcome has limited the adoption of NHB donor programmes to a relatively small number of transplant centers. Some of these have used machine perfusion preservation to try and reduce the risk of PNF, although direct evidence as to its effectiveness is lacking .

In Newcastle upon-Tyne, the introduction of machine perfusion coincided with a fall in the incidence of PNF from 54% to 7% (Balupuri et al., 2000a, b). This most probably reflects the ability to perform viability testing, and in Newcastle and other centers, kidneys are discarded if during machine perfusion intrarenal resistance measurements or levels of S-transferase are abnormal (Balupuri et al., 2000a, b; Daemen et al., 1997).

DGF occurs more frequently in NHBD (50%–100%) (Schlumpf et al., 1996) than in HBD (20%–60%) kidneys (Light et al., 2000). This increased DGF rate in NHBD kidney is likely to be a consequence of acute tubular necrosis (ATN) secondary to warm ischemic damage. Not all DGF is due to ATN, although ATN is the most common cause and it is not therefore surprising that NHBD kidneys have a higher incidence of ATN than those from HBD (Castelao et al., 1993).

Brook et al. (2004) showed in controlled donors 48% of DGF in both NHBD and HBDs, while the uncontrolled donors had a significantly higher rate (88%). It is important to note that warm time is not the only variable that influences DGF; duration of pretransplant dialysis and recipient body weight also correlate with posttransplant early graft function (Yokoyama et al., 1996).

The influence of DGF to kidney allograft survival is controversial. A number of studies have demonstrated poorer survival in HBDs with DGF (Yokoyama et al., 1994; Pfaff et al., 1998; Feldman et al., 1996) compared with HBDs with immediate function, while other authors have reported no such effect, independent of acute rejection (Troppman et al., 1996; Marcen et al., 1998).

Brook et al. (2004) demonstrate that serum creatinine is stable in recipients of kidneys from NHBDs from 3 months to 7 yr posttransplant. There is disagreement in the literature over NHBD posttransplant renal function. In some studies, NHBD kidneys achieved early serum creatinine levels in the normal range (Butterworth et al., 1997; Varty et al., 1994; Cho et al., 1998; Hoshinaga et al., 1995; Guillard et al., 1993), while other studies reported poor graft function than HBD in both the short and medium term (Casavilla et al., 1995; Wijnen et al., 1995).

NHBDS kidneys in the medium term achieve a good level of renal function with a mean serum creatinine at 12 months of 174 μmol/L (Kootstra et al., 1991) and a median of 199 μmol/L at 18 months (Varty et al., 1994). So far, there is no evidence of accelerated deterioration of NHBD kidneys because of a reduced functioning glomerular mass due to initial ischemic damage.

Acute rejection is thought to occur more frequently and with greater severity in kidneys with prolonged ischemia and DGF. This may be a result of the increased rate of detection of subclinical rejection because biopsies are taken more frequently in the presence of DGF (Nicholson et al., 1996).

Brook et al. (2003a,b) demonstrated for the first time that the high rate of DGF associated with renal transplantation from NHBDs does not lead to poor graft survival when compared with HBDs with DGF.

In recent years, there has been a reevaluation of the use of NHBDs for renal transplantation. While some studies have shown poorer graft survival for NHBD kidneys (Yokoyama et al., 1994; Feldman et al., 1996), others have demonstrated favorable graft survival compared with HBDs (Nicholson et al., 1997; Wijnen et al., 1995; Gonzalez-Segura et al., 1998). This is despite the detrimental effects of warm ischemic damage in NHBDs with consequent high rates of DGF.

Brook et al. (2003a,b) reported that despite the detrimental effects of long WIT in NHBD with consequent high rates of PNF and DGF, there have been favorable comparisons with HBD in terms of graft survival also demonstrating that NHBD kidneys display parity with HBD meeting the British Transplant Society guidelines (1998) for HBD allograft survival at 1 and 5 yr of 80% and 60%, respectively.

It would seem that long ischemic time causes reversible graft problems at an early stage in terms of PNF and DGF. Later, NHBD grafts appear to perform at least as well as those from HBDs.

Finally, as previously published, there are many ways to make transplantation demand compatible with organ supply (Chaib et al., 2007; Chaib and Massad, 2005; Chaib et al., 1995; Chaib et al., 1994).

We have always to bear in mind that the gap between demand and supply of organs for transplantation continues to grow. One solution to this problem has been to return to the practice of using grafts from NHBDs. The challenge of NHBD transplantation is to minimize the first period of warm ischemia and the consequent reperfusion injury.

Table 7.7 Non—Heart-Beating Donors—Maastricht classification (Kootstra et al., 1995a,b).

Description		
I	Dead on arrival	Uncontrolled
II	Unsuccessful resuscitation	Uncontrolled
III	Awaiting cardiac arrest	Controlled
IV	Cardiac arrest while brain dead	Controlled

Finally, NHBD kidney transplants are associated with allograft dysfunction as PNF and DGF which is related to primary warm ischemic injury. This warm ischemia is more deleterious in uncontrolled than controlled NHBDs.

Kidney transplant from NHBDs can be performed successfully. The significant degree of warm ischemic injury suffered by NHBD kidneys leads to a high incidence of DGF but the data available so far suggest that this does not adversely influence long-term graft survival (Tables 7.7—7.10).

Table 7.8 Criteria for exclusion of non—heart-beating donors (Brook et al., 2003a,b).

1	Cardiac and circulatory arrest does not last longer than 40 min.
2	The patient is between 16 and 60 years old.
3	The patient does not belong to a high risk group for immunodeficiency virus (HIV) or hepatitis B or C infection. There should be no signs of intravenous drug abuse.
4	The patient has no history of primary kidney disease, uncontrolled hypertension, or complicated insulin-induced diabetes mellitus (IDDM). There are no signs of intravascular coagulation with anuria and no signs of malignancy other than a primary (nonmetastatic) cerebral tumor.
5	There are no signs of sepsis or serious infection.
6	Patients who have died after assisted suicide or euthanasia are excluded from some protocols.

Table 7.9 Criteria for brainstem death apply to heart-beating donors (Brook et al., 2003a,b).

1	The underlying pathologic lesion should be understood.
2	There should be no pharmacologic, metabolic, or hormonal influence.
3	Pupillary, corneal, oculocephalic, vestibuloocular, and gag reflexes should be absent.
4	No pain response to stimulation in the distribution of the fifth cranial nerve.
5	A rebreathing test with 100% oxygen should be delivered to maintain satisfactory oxygenation, while ventilation is switched off. The rise in arterial pCO_2 should not stimulate respiration.

These tests are performed by two experienced clinicians on two separate occasions.

Table 7.10 Criteria for cardiac death apply for non—heart-beating donors (Brook et al., 2003a,b).

1	Deep coma
2	Absence of pulse
3	ECG evidence of asystole

Cardiac death in the context of potential organ donation is defined as occurring after 30 min of unsuccessful cardiopulmonary resuscitation under hospital conditions. Resuscitation must include external cardiac massage, intubation, ventilation, defibrillation (if indicate), and appropriate intravenous medication. Unsuccessful means that these measures did not achieve spontaneous contractile cardiac activity or peripheral circulation.

Living-donor liver transplantation

8.1 POTENTIAL EFFECT OF USING ABO-COMPATIBLE LIVING-DONOR LIVER TRANSPLANTATION

Introduction

The introduction of living-donor liver transplantation (LDLT) has been one of the most remarkable steps in the field of liver transplantation (OLT) by substantially expanding the small donor pool in countries in which the growing demands of organs are not met by the shortage of available cadaveric grafts (Raia et al., 1989). In Brazil, OLT was first performed at the University of São Paulo School of Medicine. Since then, the waiting list for liver transplantation has increased at a rate of 150 new patients per month. From 1988 to 2004, OLT increased by 1.84-fold, from 160 to 295 procedures. However, the number of patients on the liver waiting list has increased by 2.71-fold, from 553 to 1500. Consequently, the number of patients who died while on the liver waiting list has increased 2.09-fold, from 321 to 671 (Chaib and Massad, 2008a,b,c).

Worldwide, the disparity between the number of available cadaveric organs and the number of patients on the waiting list increases each year (Austin et al., 2007; Wojcicki et al., 2007). In a previous study, we projected the size of the waiting list in São Paulo State compared with the number of transplantations performed in the same period. We demonstrated that the list grows at a rate much higher than the number of transplantation procedures performed. Recently, we noted that in 2003, 4.7 OLTs were performed per million population; however, the number of OLTs performed should have been at least 33 per million population (Chaib and Massad, 2008a,b,c). We must improve the number of OLTs in our state; all sources of organ donation must be used including donors after cardiac death (Muiesan et al., 2006; Chaib, 2008; Chaib and Massad, 2008a,b,c), split-liver transplantations (Chaib et al., 2007), and routine LDLT in patients with acute and chronic liver failure. Because we do

Mathematical Approaches to Liver Transplantation. https://doi.org/10.1016/B978-0-12-817436-4.00008-4

not routinely perform LDLT, the objective of this study was to analyze, using a mathematical equation, the potential effect of using ABO-compatible LDLT in both our OLT program and waiting list in the São Paulo State, Brazil.

Materials and methods

As in Chaib and Massad (2005), we collected official data from State Center of Transplantation—State Secretariat of São Paulo about our liver transplantation program between July 1997 and October 2004.

The data related to the actual number of liver transplantation, *Tr*, the incidence of new patients in the list, *I*, and the number of patients who died in the waiting list, *D*, in the State of São Paulo since 1997 can be seen in Table 8.1.

We take the data from Table 8.1 and fitted continuous curves by the method of maximum likelihood (Hoel, 1984), to project the number of transplantations, *Tr*, incidence on the list, *I*, and deaths in the list, *D*, in future time. The resulting equations and Figs. 8.1–8.3 are as follows:

$$Tr = 107.07 \ln(\text{time}) + 72.943 \tag{8.1}$$

$$I = 553 \exp(0.17 \times \text{time}) \tag{8.2}$$

$$D = 321 \exp(0.123 \times \text{time}) \tag{8.3}$$

We next projected the size of the waiting list, *L*, by taking into account the incidence of new patients per year, *I*, the number of transplantations carried out in that year, *Tr*, and the number of patients who died in the waiting list, *D*. The dynamics of the waiting list is given by the same difference equation as in Chapter 5, $L_{t+1} = L_t + I_t - D_t - Tr_t$, that is, the list size at time $t+1$

Table 8.1 Actual number of liver transplantation, *Tr*, the incidence of new patients in the list, *I*, and the number of patients who died in the waiting list, *D*, in the State of São Paulo since 1997.

Year	Time	Tr	I	D
1997	1	63		
1998	2	160	553	321
1999	3	188	923	414
2000	4	238	1074	548
2001	5	244	1248	604
2002	6	242	1486	725
2003	7	289	1564	723
2004	8	295	1500	671

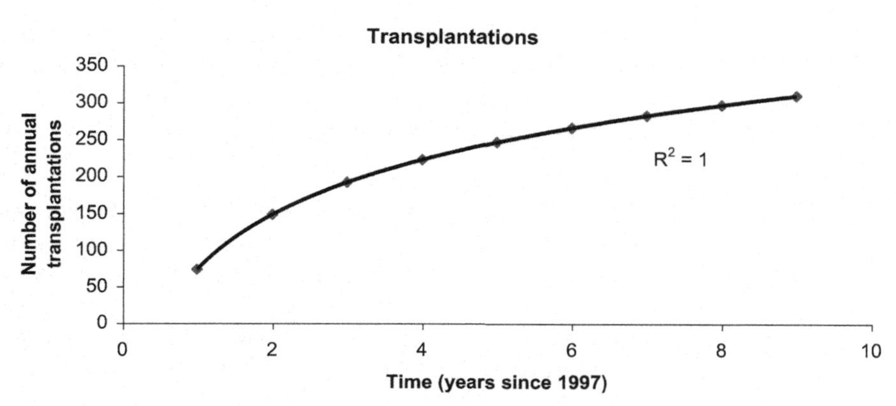

FIGURE 8.1 Annual transplantations (*diamonds*) and the fitted projection curve (*continuous line*).

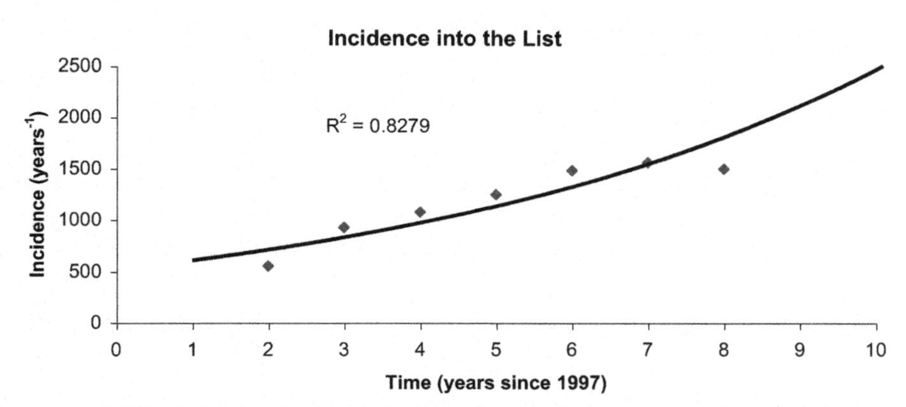

FIGURE 8.2 Annual incidence into the list (*diamonds*) and the fitted projection curve (*continuous line*).

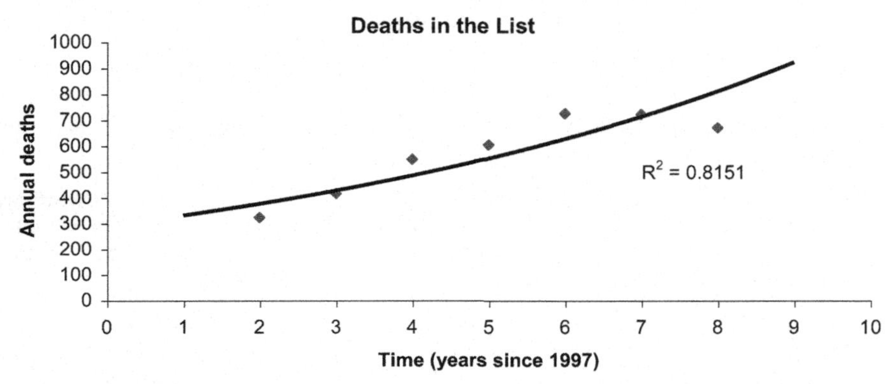

FIGURE 8.3 Annual number of deaths in the list (*diamonds*) and the fitted projection curve (*continuous line*).

is equal to the size of the list at the time t, plus the new patients getting into the list at time t, minus those patients who died in the waiting list at time t, and minus those patients who received a graft at time t. The variables I and D from 2004 onward were projected by fitting an equation by maximum likelihood, in the same way that we did for Tr.

Next we calculated the probability that a given recipient has a compatible parent. For this calculation, we first needed the frequency distribution of each blood type in the population of the State of São Paulo. Using data from Baiochi et al. (2007), we calculated the frequency distribution of each blood group allele by applying the extended equation (Bulmer, 1985) for three alleles:

$$(p+q+r)^2 = p^2 + q^2 + r^2 + 2pq + 2pr + 2qr \tag{8.4}$$

Results

The frequency distribution of each blood group and the probability of having a compatible parent are given in Table 8.2.

Assuming that, on average, 39% of the waiting list is composed of children, who should receive a liver from a parent, and 61% is composed of adults (Azeka et al., 2009), who should receive a liver from a sibling, we calculated the effect of LDLT on the waiting list using the following equation:

$$\begin{aligned} L_{t+1} = {} & L_t + I_t - D_t - [(0.51 \times 0.15) + (0.32 \times 0.17) + (0.13 \times 0.10.6) \\ & +(0.04 \times 0.03)] \times 0.39 \times L_t - [(0.51 \times 0.35) + (0.32 \times 0.12) + (0.13 \times 0.05) \\ & +(0.04 \times 0.03)] \times 0.61 \times L_t \end{aligned} \tag{8.5}$$

The numerical term in Eq. (8.5) represents the number of transplantations performed using parental donations. The effect of an LDLT program is shown in Fig. 8.4. Note that because of the relatively low probability of

Table 8.2 Frequency distribution of blood groups and probability of compatible parents or siblings in the State of São Paulo, Brazil.

Blood group	Prevalence in the population	Probability of having a compatible parent	Probability of having at least one compatible sibling
O	0.5067	0.154	0.350
A	0.3217	0.169	0.122
B	0.1345	0.059	0.052
AB	0.0371	0.029	0.029

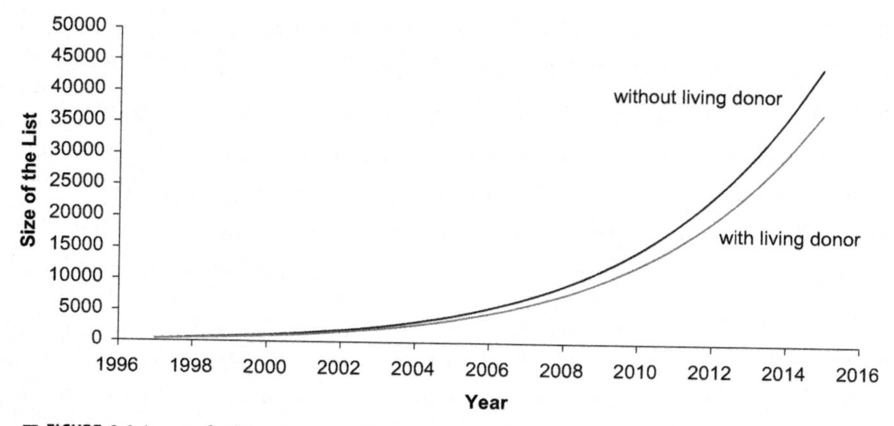

■ **FIGURE 8.4** Impact of a living-donor transplantation program on the waiting list of the State of São Paulo, Brazil.

parent or sibling compatibility, the effect of such a program would be a reduction of the waiting list of about 16.5% at 20 years after its introduction. However, an estimated 3600 lives could be saved in the same period.

Conclusions

The inequality between the supply of grafts and the demand for transplantation has forced the transplantation community to devise ways to increase the number of available livers for transplantation such as use of extended-criteria donor grafts and living-donor grafts. In the United States, the number of patients awaiting liver transplantation greatly exceeds the supply of cadaveric organs. More than 15,000 patients are currently registered on the liver transplantation waiting list of the United Network for Organ Sharing, and only about 4500 cadaveric livers become available for transplantation each year. Cadaveric organ availability seems to have reached a plateau despite many initiatives intended to increase organ donation.

The waiting time for liver transplantation has increased steadily each year, rising from approximately 1 month in 1988 to more than 1 year in 1999. Particularly in the State of São Paulo, Brazil, the problem is much the same; there is a critical shortage of donor organs, a problem that will worsen in the foreseeable future (Chaib and Massad, 2008a,b,c). The introduction of LDLT has been one of the most remarkable steps in the field of OLT. First introduced for use in children in 1989 (Raia et al., 1989), its adoption for use in adults followed only 10 years later.

As the demand for OLT continues to increase, LDLT, despite its complications, provides life-saving therapy for many patients who would otherwise die awaiting a cadaveric organ. In recent years, LDLT has been shown to be a clinically safe addition to deceased-donor liver transplantation and has substantially extended the limited donor pool. As long as the donor shortage continues to increase, LDLT will have an important role in the future of OLT (Nadalin et al., 2006).

In children, LDLT is a popular option for liver transplantation. Adult-to-child LDLT using the left lateral portion of the liver is safe and effective. The procedure has helped to reduce the number of children who die awaiting liver transplantation. It also has helped avert family disruption caused by having a child and at least one parent away from home, staying near the transplantation center while awaiting a chance for a cadaveric organ transplantation.

Although ABO-incompatible LDLT is feasible (Testa et al., 2008), the results are poorer than with compatible grafts. LDLT still accounts for less than 5% of adult liver transplantations, substantially less than in kidney transplantation, in which LDLT accounts for approximately 40% of all transplantations performed.

In this chapter, we used a mathematical equation to analyze the potential effect of ABO-compatible LDLT on liver transplantation and the waiting list in São Paulo, Brazil, considering only first-degree sibling (adult-to-adult) LDLT and the possibility that at least one parent would donate a partial graft to their child. Because of the relatively low probability of parent or sibling ABO-compatibility, the effect of such a program would be a reduction of the waiting list of about 16.5% at 20 years after its introduction. However, an estimated 3600 lives could be saved in the same period.

We also attempted to calculate the effect of a nonrelated donor on the liver waiting list. Considering that each recipient would find an ABO-compatible donor, we estimated that for a recipient with blood group type O, the probability is 50.6%; for a recipient with type A is 82.84%; for a recipient with type B is 64.12%; and for a recipient with type AB is 100%. We assumed that those potential donors would agree to donate part of their liver. If we calculate the total effect of a nonrelative donation on our liver waiting list, we predict that unless the probability of donation is greater than 10%, the effect on our liver waiting list would be negligible. Thus, we did not include these calculations in our results. We did not consider rhesus factor in our model because this is not a relevant limitation for donation in Brazil.

A word of caution about our results is necessary. According to the literature (Gruessner and Benedetti, 2008), only 11%–40% of all potential donors qualify for live donation because of medical, psychosocial, or anatomical exclusion factors. Therefore, our results represent an upper limit to the number of saved lives.

In conclusion, use of adult-to-adult and adult-to-child LDLT would reduce the pressure on our OLT waiting list by reducing the size of the waiting list by at least 16.5% at 20 years after its introduction. Such an LDLT program in São Paulo State could save an estimated 3600 lives over the same period.

The MELD scale

9.1 HISTORY OF THE MELD SCALE

Despite of increment of liver transplantation (LT), there are still growing numbers of accumulated patients in waiting list. Ultimate goal of allocation system is balancing between justice and utility, which means to optimize the use of scarce donor organ resource and to reduce waiting list mortality and furthermore to maximize long-term outcome.

Historically, the severity of cirrhotic liver disease has been calculated using the Child-Pugh (CP) class. The variables used in the calculation of the CP class were not the result of systematic analysis but rather emerged from clinical experience. The CP class has been shown to be valuable in determining prognosis in cirrhotic patients undergoing medical management (Christensen et al., 1984). However, other authors have reported the Child classification and Pugh score failed to predict postoperative 30-day mortality (Rice et al., 1997). This failure may be related to the limitations of the CP system, including the subjective interpretation of parameters such as ascites and encephalopathy, as well as a limited discriminatory ability.

The Model for End-Stage Liver Disease (MELD) score originally was developed and validated to assess the short-term prognosis of patients with cirrhosis undergoing the transjugular intrahepatic portosystemic shunt (i.e., TIPS) procedure (Malinchoc et al., 2000). In 2002, after validation in multiple types of liver disease patients, the MELD score was adopted as a standard by which liver transplant candidates with end-stage liver disease were prioritized for organ allocation in the United States (Kamath et al., 2000), and in 2006 in Brazil. This scoring system utilizes three widely available laboratory values—total bilirubin (g/dL), creatinine (g/dL), and international normalized ratio (INR) of the prothrombin time.

Mathematical Approaches to Liver Transplantation. https://doi.org/10.1016/B978-0-12-817436-4.00009-6

MELD formula used to calculate the severity score is as follows: 3.8 $\log_e$ (total bilirubin mg/dL) +11.2 $\log_e$ (international normalized ratio [INR]) +9.6 $\log_e$ (creatinine mg/dL) +6.4.

The MELD system had an immediate impact on the liver transplant landscape leading to a reduction in the number of wait list registrants for the first time ever, and a 15% reduction in mortality among those on the waiting list (Freeman et al., 2004a,b,c; Fink et al., 2005). Since the introduction of MELD as the primary allocation system, there has always been a goal of improving this mathematical prioritization model (Wiesner et al., 2003a,b).

The properties of an ideal allocation score for liver transplant remain an open question, and there is no current consensus in the transplant community. There is a significant shortage of donor liver allografts, with 14,413 patients on the wait list for liver transplant and only 7841 liver transplants performed in 2016 (Kim et al., 2008; Nagai et al., 2018). With priority on transplant wait lists determined by the MELD score, a desirable use for the MELD score would be to define a cutoff above which liver transplant would confer a survival benefit. Furthermore, when donor livers are allocated based on MELD, there should be correspondence between higher MELD scores and greater benefit. To fulfill the principle of organ allocation according to medical need, the ideal allocation score would be able to define a relatively narrow group of patients, that is, one similar in size to the number of available donors, who would benefit most from receiving a liver transplant.

Prior studies have examined variations in transplant benefits, defined as the difference in survival posttransplant versus that shown if the patient had remained on the wait list. In a 2005 study, patients with MELD scores between 18 and 20 had a 38% reduction in mortality risk after transplant, whereas patients with MELD scores between 15 and 17 had 21% greater risk of mortality after transplant. Such comparisons highlighted the evident lack of transplant benefit for patients with low MELD (Nagai et al., 2018). However, after a threshold of transplant benefit was described using the original MELD score, serum sodium was added to the MELD score to improve prediction of survival.

Kim et al. (2008) proposed a modified scoring system to further decrease wait list mortality that incorporates serum sodium concentration into the MELD equation (MELD-Na). This MELD-Na score was officially implemented for liver graft allocation in January 2016.

The United Network for Organ Sharing (UNOS) implemented the resulting MELD-Na score, using the following formula:

$$\text{MELD} - \text{Na} = \text{MELD} + 1.32 \times (137 - \text{Na}) - [0.033 \times \text{MELD} \times (137 - \text{Na})]$$

Liver allocation based on MELD-Na score successfully improved wait list outcomes and provided significant benefit to hyponatremic patients.

9.2 THE IMPACT OF THE INTRODUCTION OF MELD ON THE DYNAMICS OF THE LIVER TRANSPLANTATION WAITING LIST IN SÃO PAULO, BRAZIL

Introduction

The global liver allocation system in use until 2002 was based on the Child-Turcotte-Pugh (CTP) scale and other scores as well as on the waiting time; the system became the major discriminator of patients on the wait list without reflecting their actual liver dysfunction. The system for prioritizing adult patients on the wait list in the United States has since changed from a status-based algorithm using the CTP scale to a system using a continuous MELD (Freeman et al., 2004a,b,c). The MELD score was originally developed to predict the survival after a TIPS procedure (Malinchoc et al., 2000).

In São Paulo, the time on the waiting list was the primary criterion adopted to allocate deceased donor livers until July 15, 2006. After this date, MELD was the basis for the allocation of deceased donor livers for adult transplantation. The MELD score primarily sought to increase access to transplantation for severely ill patients as a means to reduce the mortality rate of the waiting list patients. The MELD score does not consider the posttransplant benefit.

Our aim was to compare the wait list dynamics in the pre-MELD (1997–2005) and post-MELD (2006–12) periods in the state of São Paulo, Brazil.

Material and methods

A retrospective study was initially conducted that included the waiting list data of all the liver transplant candidates from July 1997 to December 2012 in the state of São Paulo, Brazil. The data were from the liver transplant research database of the Health Secretariat of São Paulo.

In this study, inclusion was restricted to adult patients (>18 years) who were candidates for LT. Living donors' related LT cases and split livers and pediatric recipients were excluded from the study. The candidates were divided into the pre-MELD group for those listed from July 2007 to December 2005 and the post-MELD group for those listed from July 2006 to December 2012. The patients with hepatocellular carcinoma according to the Milan criteria were granted additional MELD points. On the first year after the implementation of MELD, these patients arbitrarily received a MELD score of 29, which was changed to a MELD score of 24 in the second year after the implementation of MELD.

The patients who had acute hepatic failure and were not allocated by the MELD system were excluded. The Organ Procurement Organization obtains the waiting list additions, modifications, and removals directly from the transplant centers. Patients were removed from the wait list for the following major reasons: clinical improvement without a transplant; death; or being unable to receive a transplant based on the health status of the patient.

MELD was calculated at the time of transplantation, as previously described (Chaib et al., 2013a,b), using the following objective variables: the serum creatinine, bilirubin, and INR levels. The minimum acceptable value for INR, creatinine, and bilirubin is 1. The maximum acceptable value for serum creatinine is 4 mg/dL. If the patient had been dialyzed twice within 7 days, then the value for the serum creatinine would be 4. The maximum value for the MELD score is 40. The laboratory data and the MELD score were not collected in the pre-MELD era. For those who received a transplant after the implementation of the MELD system, the MELD score at transplantation was used as a marker of severe liver disease.

The data related to the actual number of LTs, the incidence of new patients on the list (Chaib and Massad, 2005), and the number of patients who died while on the wait list from 1997 to 2005 (the pre-MELD era) and from 2006 to 2012 (the post-MELD era) are shown in Tables 9.1 and 9.2, respectively.

We used the data from Table 9.1 (1997–2005) and Table 9.2 (2006–12) fitting a continuous curve by the maximum likelihood method (Hoel, 1984) to project the number of future transplantations (see Chaib and Massad, 2005). The pre-MELD era wait list dynamics were previously described

Table 9.1 Actual number of liver transplantation (1997–2005) (Tr), the incidence of new patients in the list (I), and the number of patients who died in the waiting list (D), in the state of São Paulo since 1997 until 2005 (Pre-MELD).

Year	Tr	I	D
1997	63	–	–
1998	160	553	321
1999	188	923	414
2000	238	1074	548
2001	244	1248	604
2002	242	1486	725
2003	289	1564	723
2004	295	1500	671
2005	299	1907	662

Table 9.2 Actual number of liver transplantation (2006–12) (Tr), the incidence of new patients in the list (I), and the number of patients who died in the waiting list (D), in the state of São Paulo since 2006 until 2012 (Post-MELD).

Year	Tr	I	D
2006	349	1566	895
2007	330	1022	734
2008	454	1213	490
2009	609	1287	455
2010	671	1415	403
2011	609	1577	470
2012	501	1488	441

by our group (Chaib and Massad, 205). The resultant equations are as follows:

$$Tr = 107.07(\text{year}) + 72.943, \quad Tr = 500(\text{year}) + 184.7 \qquad (9.1)$$

or the first (1997–2005) and second (2006–12) groups, respectively.

Results

The results are visualized in Figs. 9.1 and 9.2, in which the number of transplants performed is fitted to the above function used to project the list size.

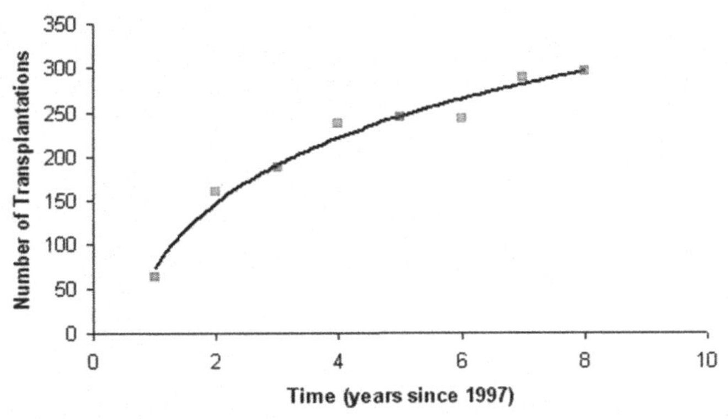

■ **FIGURE 9.1** Fitting curve by the method of maximum likelihood to data from Table 9.1 to project the number of transplantations, *Tr*, in future time (pre-MELD).

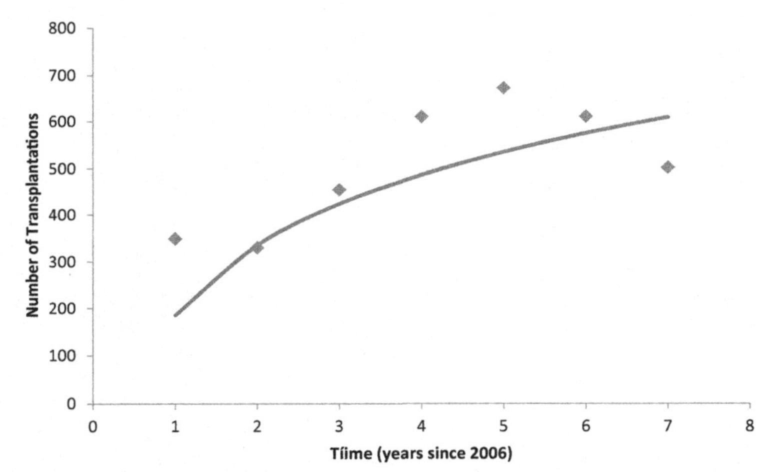

■ **FIGURE 9.2** Fitting curve by the method of maximum likelihood to date from Table 9.2 to project the number of transplantations, *Tr*, in future time (post-MELD).

The number of transplants from 1997 to 2005 and from 2006 to 2012 increased nonlinearly, with a clear trend to leveling to equilibrium of approximately 350 and 500 cases per year, respectively.

We projected the size of the waiting list by considering the incidence of new patients per year, the number of transplants performed in that year, and the number of patients who died while being on the waiting list. The dynamics of the waiting list is presented by the difference Eq. (4.1), $L_{t+1} = L_t + I_t - D_t - Tr_1$, with the list size at time being equal to the size of the list at time plus the new patients added to the list at time minus those patients who died while on the waiting list at time and minus those patients who received a graft at time. The variables from 2006 were projected by fitting an equation by the maximum likelihood, in the identical manner as the equation was fitted for.

The introduction of the MELD score had a significant effect on the wait list dynamics in the first 4 years after its introduction; however, the curves diverge from there, implying a null long-range effect by the MELD scores on the wait list (Figs. 9.3 and 9.4).

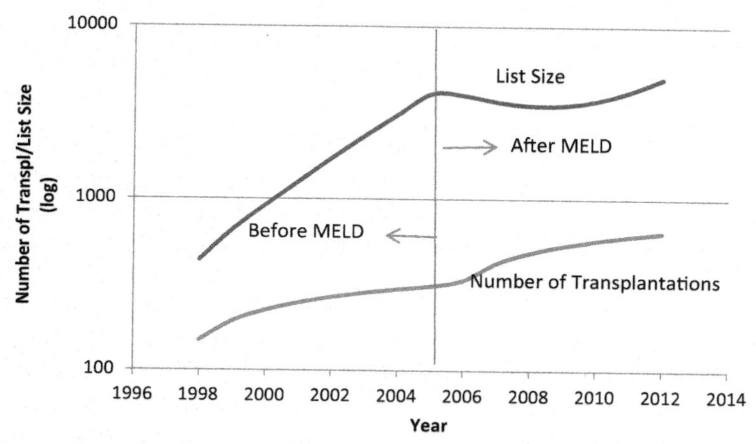

■ **FIGURE 9.3** Fitting of Eq. (9.1) on data from Tables 9.1 (pre-MELD) and 9.2 (post-MELD).

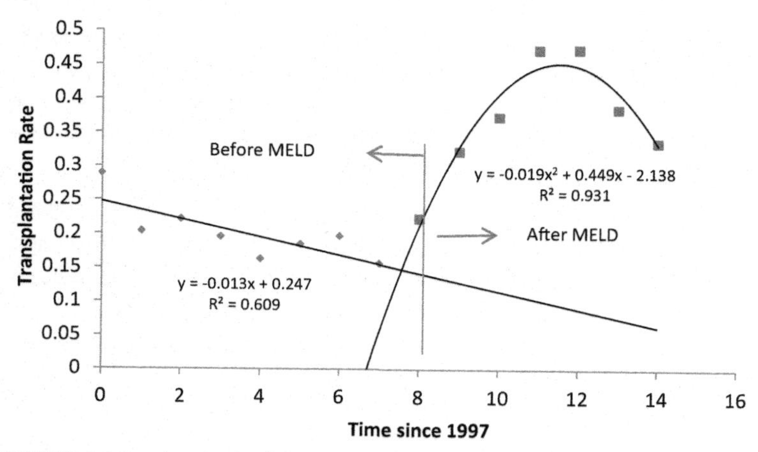

■ **FIGURE 9.4** Transplantation rate (proportion of those patients transplanted with respect to those who enter the waiting list per time unit), comparing the pre- and post-MELD periods.

Conclusions

The increased mortality of patients waiting for a liver transplant and the shortage of donor organs induced efforts to improve the allocation criteria for LT candidates. The introduction of the MELD system in the United States for graft allocation resulted in a 3.5% reduction in wait list mortality, whereas the early-stage survival of liver transplant recipients remained unchanged, despite the selection of more seriously ill patients for transplantation (Freeman et al., 2004a,b,c).

Although MELD eliminates subjective assessments and shows accuracy for predicting the outcome in patients with decompensated cirrhosis, it has several limitations (Voigt et al., 2004; Wiesner et al., 2006). One of the limitations of the MELD score is that the components of the MELD score were found to independently and individually predict death on the wait list.

The major reason for MELD implementation was to decrease the number of deaths of wait list patients, providing each patient an identical probability of receiving a transplant at presumed fixed condition levels. Previously, priority was determined by a more complex system, in which the waiting list time and patient condition, classified in a semiquantitative way, were linked (the presence of encephalopathy and ascites as well as the waiting time and patient location). An ultimate goal has been to end the privilege of selecting the candidate on a clinical basis (considering various parameters such as the primary disease, degree of residual liver function, extrahepatic involvement, waiting list time, and donor-related risk), which was once a prerogative of the transplant surgeon.

The role of the match between the "donor quality" and the severity of the recipient's disease has not been completely investigated. In many transplant centers, standard livers are routinely transplanted in low-risk patients, whereas marginal donors are reserved for high-risk patients. Our group suggested that better results are obtained if the risk related to the donor and the risk related to the recipient are not merged.

The effectiveness of MELD as a prognostic index has been fully validated in cirrhotic patients waiting for transplantation. The role of MELD as a prognostic index in liver transplant patients is controversial. The efficacy of MELD in predicting graft survival has been reported in several uncontrolled single center cohorts characterized by an intrinsic high-risk condition (Briceno et al., 2005).

A major challenge facing the field of LT is the critical shortage of donor organs, which has led to a dramatic increase in the number of patients on the wait list as well as in the waiting time of the patients. In the pre-MELD era, the number of LTs increased 1.86-fold (from 160 to 299) from 1998 to 2005; however, the number of patients on the liver wait list increased 3.44-fold (from 553 to 1907). The number of deaths of the wait list patients increased 2.06-fold (from 321 to 662, Fig. 9.1).

The implementation of the new liver allocation system in our state has required a change in the disease severity score, with minimal weighting being allocated to the waiting time compared with the previous system that was based on the CTP score and on the waiting time.

The lower priority placed on the waiting time has improved organ access for worse transplant candidates, and worse patients were selected from the pre-transplant waiting list. This fact is reflected by the significant increase of the median MELD score at the time of LT as well as by the decreased median waiting time. We found that the median time on the wait list decreased only for the patients in whom LT was performed, whereas a significant proportion of patients with lower MELD scores are likely to have much longer waiting times.

After the implementation of MELD, we observed that the number of liver transplants increased 1.43-fold (from 349 to 501) from 2006 to 2012; the number of patients on the LT wait list was slightly reduced (0.95-fold), from 1566 to 1488 patients. The number of deaths of wait listed patients has been significantly reduced (2.02-fold), from 895 to 441 patients (Fig. 9.2).

The major controversy following the implementation of MELD is the balance in organ allocation between reduced wait list mortality and the best posttransplantation outcome (Briceno et al., 2005). Numerous studies have investigated, with varying results, the prognostic value of the MELD score for early and late posttransplant survival (Ahmad et al., 2007; Brandão et al., 2009; Chaib et al., 2013a,b). At our center, the recipients with a MELD score between 20 and 29 received organs fulfilling at least one extended donor criterion significantly more frequently. After the implementation of MELD, rating patients with a higher score based on longer waiting times became meaningless, and acceptance of an organ from extended criteria donors via center-based allocation represents the only opportunity for transplantation.

In conclusion, the implementation of the MELD score resulted in a shorter waiting time until LT for patients. The MELD system had a significant effect on the wait list dynamics in the first 4 years; however, the curves diverge from that point, implying a null long-term effect by the MELD scores on the list of patients waiting for transplantation.

9.3 DOES THE PATIENT SELECTION WITH MELD SCORE IMPROVE SHORT-TERM SURVIVAL IN LIVER TRANSPLANTATION?

Introduction

One of the most controversial and significant problems in the LT area is the high mortality rates on the waiting list. Recently, the United Network for Organ Sharing implemented a new policy using the MELD. The MELD score, based on serum creatinine and bilirubin levels as well as prothrombin (Cuomo et al., 2008; Stell et al., 2004), was designed to prioritize orthotopic LT for patients with the most severe liver disease rather than time spent on the waiting list. Subsequent studies proved that MELD score was really effective to decrease the mortality rate on the waiting list. The number of patients who either died on the waiting list or were removed as too sick for transplant decreased from 1220 in 2001 to 1113 in 2002. When adjusted for changes in the waiting list size, this represents a 23% decrease in deaths.

The aim of this review was to examine and discuss several aspects of the new allocation system, including short-term patient survival rate according to the MELD score comparing it with pre-MELD era.

Methods

A structured literature review at the online database PubMed/Medline/Scielo using both the English and non-English literature and also the terms of "liver transplantation and/or MELD score and/or survival rate" was performed from 2002 to 2009. A retrospective cohort study using data from the UNOS Scientific Registry and European Liver Transplant Statistic was performed. In countries within the alliance of Eurotransplant, the MELD score for prioritization of patients awaiting for LT was initiated in November 2006 and at present little information is available concerning the prognostic ability of this allocation system compared to the previous one, which was based on CTP score and waiting time.

678 articles were found. Exclusion criteria included (1) irrelevance of the subject; (2) lack of information; (3) nonsignificant number of transplantation (fewer than 100); and (4) patients who were transplanted previously were excluded for the purposes of this analysis because of the variability in the methods used for determining prothrombin time. Were also excluded those who underwent combination transplant procedures (e.g., combined liver–renal procedure) or a prior liver transplant. By the end, 30 articles have been reviewed. 1-year patient survival rate related to MELD score in those papers was analyzed. The data of pre-MELD era were collected from UNOS and European Liver Transplant Statistic. As far as to assess the short-term survival, patients with hepatocellular carcinoma were not given a priority MELD score.

Liver failure patient in pre-MELD era in United States was ranked as liver status 3, 2B, 2A, and 1 according to their liver function meaning that the worst patient was liver status 1. MELD score with higher standard deviation was excluded from statistic analysis. Patients with L1, L2A, and MELD>20 were considered the sickest, whereas those with L2B, L3, and MELD<20 were considered the healthiest. Patient survival rate from pre-MELD era was summarized and compared it with currently patient survival rate in MELD era. The X^2 test was applied for comparing both groups. Statistic significance was considered for $P < 0.05$.

Results

Analysis of patient survival rate significance between the pre-MELD (1996–2001) and post-MELD era (2003–09) (Data collected from Organ Procurement and Transplantation Network (USA) and European Liver Transplant Statistic), is shown in Table 9.3.

Discussion

Increased mortality of patients on the waiting list for LT and shortage of donor's organ gave rise to efforts to improve allocation criteria for LT candidates. The MELD was developed to predict short-term mortality in patients with cirrhosis. The model's accuracy to predict short-term mortality among patients with end-stage liver disease has been largely established. It has since become the standard tool to prioritize patients for LT.

Since the implementation of MELD on liver allocation in the United States, new registrations on the waiting list have been reduced and transplantation rates have improved without increasing mortality rates of waiting candidates or changing early transplant rates. Some authors showed no correlation between MELD and short-term posttransplant survival.

Table 9.3 Analysis of patient survival rate significance between the pre-MELD (1996–2001) and post-MELD era (2003–09) (Data collected from Organ Procurement and Transplantation Network (USA) and European Liver Transplant Statistic)).

Group		Gravity status	America		Europe	
			PSR (%)	N Alive	PSR (%)	N Alive
Pre-MELD era	I[a]	E1	79.6	1153	—	—
		E2A	82.0	192	—	—
	II	E2B	89.1	352	—	—
		E3	93.3	96	—	—
114 Post-MELD era	III[a]	≥ 20	88.3	5914	66.5	114
	IV	≤ 20	87.4	10,333	84.4	912

(—) Data not available. L, liver status; MELD, Model for End-stage Liver Disease; N, number of patients; PSR, patient survival rate.
[a]Group I (L1 and L2A) × Group III (MELD $\geq$ 20)—significant ($p < 0.0001$); Group II (L28 and L3) × Group IV (MELD $\leq$ 20)—not significant; L1, L2A, and MELD $\geq$ 20—the sickest patient; L2B, L3, and MELD $\leq$ 20—the healthiest patient.

However, other reports suggested that pretransplant MELD predicted posttransplant survival (Biggins et al., 2006).

The authors, in a retrospective study, evaluated the capacity of the MELD score, at the time of LT, to predict posttransplant survival and found that MELD score had significantly better prediction in the outcome of LT in sickest patient (MELD>20) when compared with healthiest patient (MELD<20). Not only patients' survival rate but also death on the waiting list and well removal from the waiting list due to poor condition were analyzed; additionally, was extended to the patient observation time over a period of 1 year. Our data do not argue against the use of MELD concerning prioritization of patients during initial period on the waiting list. But in patients with a longer time on waiting list, other methods of gravity evaluation (such as CTP) may be used for assessment of patient prognosis. Some parameters that are not used to calculate the MELD score were also identified: retransplantation and the need for mechanical ventilation. The impact of including patient need for pretransplant mechanical ventilation into MELD-based allocation is unknown. Intuitively, it is expected that patients requiring ventilator support are likely to have an increased risk of death without rapid transplantation. Similarly, whether the patient is awaiting a primary graft or a retransplant was not considered in the development of the MELD scoring system. Numerous studies identify retransplantation procedures to have significantly worse outcome than primary transplantations (Facciuto et al., 2000). Whether patients awaiting a retransplant also have reduced survival compared with patients waiting for a first transplant remains to be determined. It is even unclear whether the MELD score has comparable predictive capacity in this unique and difficult subgroup of patients. Inclusion of either of these variables into MELD may improve allocation to patients most in need; however, the efficient use of cadaver livers may be decreased because each also portends inferior outcome posttransplant.

Recent reports have shown that patients with higher MELD scores have poorer posttransplant survival, but the correlation between MELD and posttransplant patient survival was only marginally better than that for graft survival; the MELD score seems less predictive than the specific disease (Roberts et al., 2004). Although this is not surprising, the MELD score was designed to predict pretransplant survival. The characteristics that are predictive of survival while awaiting transplantation are not identical to those that substantially contribute to success of a liver transplant and predict posttransplant survival (age, specific disease, surgical times, immunosuppressive therapy, etc.). However, these differences indicate that our understanding of the predictors of transplant success and failure remains dynamic and that the reevaluation of risk factors and survival predictors is an important ongoing activity (Freeman et al., 2004a,b,c). Using a combination of MELD with other pre- or posttransplant factors may be a better alternative.

In countries within the alliance of Eurotransplant, the MELD score for prioritization of patients awaiting LT was initiated in November 2006, and at present little information is available concerning the prognostic ability of this allocation system compared to the previous system which was based on CTP score and waiting time, that is why there is a gap on Table 9.1 in data of pre-MELD era in Europe.

The increasing numbers of standard exceptions for MELD scores, for example cholestatic diseases, reflect the clinical need to improve this allocation system. Although the authors' study does not argue against the use of the MELD score for short-term allocation of organs and prioritization of recipients, the long-term prediction of mortality or removal from waiting list in patients awaiting LT might be better assessed by CTP score than the MELD.

Although overall outcomes of patients whose MELD scores were high at the time of LT were inferior to those of patients whose MELD scores were lower, there was no significant difference for specific thresholds of MELD above which LT was discouraged and a patient should be removed from the waiting list.

Finally, the MELD score has significantly improved short-term patient survival rate for the sickest patient on the waiting list for LT; additionally, it does not have any significant impact in patient survival rate for the healthiest patient. It is fair to say that the impact of pretransplant MELD is maximal during the first year posttransplant; however, better predictive models are needed to assess the survival benefit with LT.

Liver tumors and liver transplantation

10.1 A MATHEMATICAL MODEL FOR OPTIMIZING THE INDICATIONS OF LIVER TRANSPLANTATION IN PATIENTS WITH HEPATOCELLULAR CARCINOMA

Introduction

As mentioned in the previous chapters, liver failure occurs when large parts of the liver become damaged beyond repair, and the liver is no longer able to function. It may be caused by infections, toxic substances, inherited diseases, or malnutrition (Chaib et al., 2012). The chronic aggression of liver tissue by one of the causes of liver failure can end up in primary hepatocellular carcinoma (HCC), a deadly condition to which liver transplantation is the only option, with variable success rate of a close-to-normal life after the surgery (Chaib and Massad, 2008a,b,c).

Within the past 5 years, the proportion of patients with HCC in waiting lists for LT has increase dramatically: this proportion has reached more than 26% across Europe and 34% in the United States (Thuluvath et al., 2010).

The Milan criteria, MC, are defined by the presence of a single nodule up to 5 cm, up to three nodules none larger than 3 cm, with no evidence of extrahepatic spread or macrovascular invasion. The countries that adopt the MC law allow patients only within MC to be evaluated and considered for LT. This policy implies that some patients with HCC slightly more advanced than those allowed by the current strict selection criteria will be excluded, even though LT for these patients might be associated with acceptable long-term outcomes (Ryckman et al., 2008; Mazzaferro et al., 2009).

We propose a mathematical approach to study the consequences of relaxing the MC for patients with HCC who do not comply with the current rules for inclusion in the transplantation candidate list. We consider overall 5-years survival rates compatible with the ones reported in the literature. We simulate our model to reproduce what is known about the survival of the two groups of patients (those who comply with the strict MC and those who

Mathematical Approaches to Liver Transplantation. https://doi.org/10.1016/B978-0-12-817436-4.00010-2

do not) and calculate the best strategy that would minimize the total mortality of the affected population, that is, the total number of people in both groups who die after 5 years of the implementation of the strategy, either by posttransplantation death or by death due to the basic HCC.

The model

We assumed, as a model, that HCC patients present themselves along a short time interval ΔT with tumors of variable sizes. We call this interval "at presentation." During this time interval, we assumed that N HCC patients are included in the transplantation waiting list, and that F livers are available to these patients.

The model is based on four assumptions, namely,

1. The mortality rate of nontransplanted, α_{nt}, and transplanted, α_t, HCC patients are described by the following ad hoc expressions:

$$\alpha_{nt}(s) = \alpha_0\left(\alpha_1 - e^{-\delta_1 s}\right) \quad (10.1)$$

and

$$\alpha_t(s) = \alpha_0 e^{\delta_2 s} \quad (10.2)$$

where δ_i $(i = 1, 2)$ are the parameters, such that $\delta_1 > \delta_2$ and s is the size of the tumor. In Eq. (20.1), when $\alpha_1 = 2$, the above mortality rates coincide for $s = 0$. Because this is necessary, we assume $\alpha_1 = 2$ for the rest of this chapter. Note that s is the size of the tumor at the moment patients get into the transplantation program. So, Eqs. (10.1) and (10.2) take into account the fact that tumors grow with time and so does the mortality rates. This is included in a rather cavalier manner in Eqs. (10.1) and (10.2) as the functional relationship of tumor growth–related mortality with time is not known.

Eqs. (10.1) and (10.2) are illustrated in Fig. 10.1, in which the mortality rates for both the transplanted and nontransplanted HCC patients as a function of the tumor size s at presentation are shown.

The probability of surviving after T years for nontransplanted and transplanted patients, $\pi_{nt}(s)$ and $\pi_t(s)$, respectively, as a function of their tumor size, s, at the time individuals are included in the transplantation program, is given by

$$\pi_{nt}(s) = \exp(-\alpha_{nt} T) \quad (10.3)$$

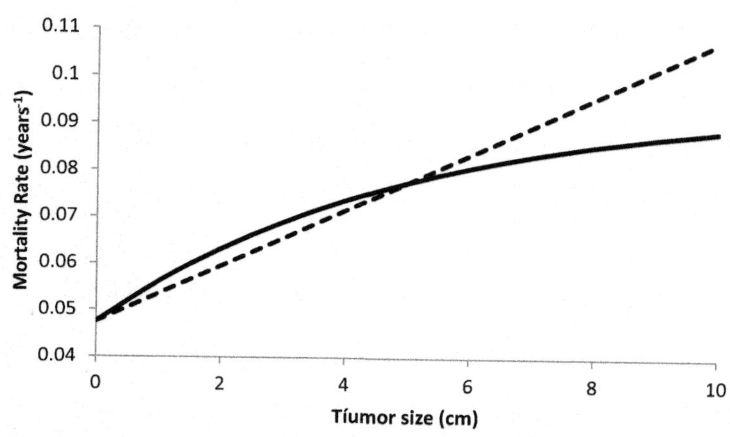

FIGURE 10.1 Mortality rates for transplanted (*dotted line*) and nontransplanted (*solid line*) HCC patients. Results of the theoretical population analyzed, according to Eqs. (10.1) and (10.2) with $\alpha_0 = 0.048$, $\delta_1 = 0.006$, and $\delta_2 = 0.2$.

and

$$\pi_t(s) = \exp(-\alpha_t T) \tag{10.4}$$

Eqs. (10.3) and (10.4) result in survival probabilities after T years, which are in agreement with data available in the literature. They were used to calculate the form and parameters of Eqs. (10.1) and (10.2).

2. The mortality of both transplanted and nontransplanted HCC patients is a monotonically increasing function of tumor size at presentation (tumor size is, therefore, taken as an indication of gravity).
3. The number of available livers to be grafted, F, is limited and always less than the total number of HCC, N, who have transplantation indication.
4. Finally, the tumor size, s, at the time individuals are included in the transplantation program, is distributed in the HCC population according to an exponential distribution, such that the probability that a given HCC patient has tumor size s is described by the *probability density function*:

$$f(s,\lambda) = \lambda e^{\lambda s} \tag{10.5}$$

where λ is the *rate parameter* of the distribution. This implies that in an HCC population, many individuals have tumor of small size and a very low number of who present tumors of larger size. Again, this distribution of tumor size is at the moment the patients get into the transplantation program. The *cumulative distribution function* is given by

$$F(s, \lambda) = \int_0^s \lambda e^{\lambda t} dt = 1 - e^{\lambda s} \tag{10.6}$$

Eq. (10.6) means the probability that a given HCC patient has tumor size equal or less than s.

The exponential distribution has mean (*expected value*) equal to

$$E[s] = \frac{1}{\lambda} \tag{10.7}$$

and variance

$$Var[s] = \frac{1}{\lambda^2} \tag{10.8}$$

1. In Fig. 10.2, we show the actual distribution of tumor size, fitted to an exponential distribution. The parameter λ in this case is equal to 0.16. As the total number of patients in these samples was 500 patients, this implies in an average size of 6.5 cm and a 95% confidence interval of [2.9; 3.7] (Yao et al., 2001a,b; Bismuth et al., 1993; Marsh et al., 2000).

With the above assumptions, we define $p(s)ds$ as the proportion of individuals with tumor size between s and $s + ds$; $x(s)ds$ as the proportion of transplanted patients with tumor size between s and $s + ds$; and $y(s)ds$ as the

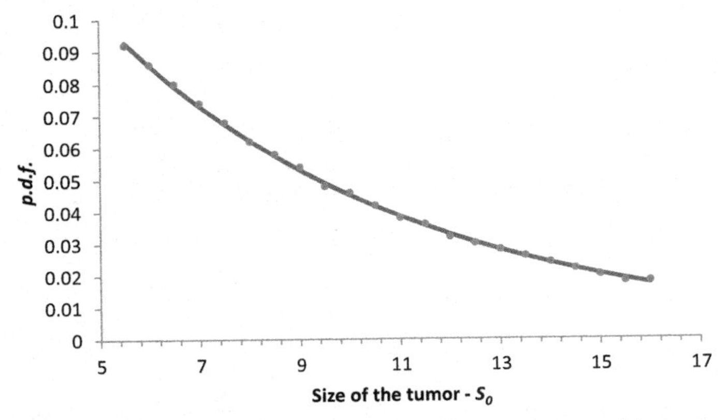

■ **FIGURE 10.2** Frequency distribution of tumor size. Dots represent actual values, and the line is the exponential fitting to the real data ($R^2 = 0.998$). Parameter $\lambda = 0.16$.

proportion of nontransplanted patients with tumor size between s and $s + ds$. These proportions are related such that

$$x(s) = \frac{F}{N}p(s) \tag{10.9}$$

and

$$y(s) = \left(1 - \frac{F}{N}\right)p(s) \tag{10.10}$$

Eqs. (10.9) and (10.10) can be interpreted as follows: a proportion $p(s)$ of the HCC patients has tumor size s, of which a fraction $\frac{F}{N}$ is transplanted, and its complement $\left(1 - \frac{F}{N}\right)$ is not transplanted, such that $x(s) + y(s) = p(s)$. Note that, this was a particular transplantation policy. For example, we could replace Eqs. (10.9) and (10.10) by $x(s) = g(s)\frac{F}{N}p(s)$ and $y(s) = \left(1 - g(s)\frac{F}{N}\right)p(s)$, where $g(s)$ is some bias toward any eventual tumor size preference. In this work, $g(s) = 1$, meaning that all HCC patients have the same chance of being transplanted (no bias). According to the MC,

$$g(s) = \begin{cases} 1 & \text{if} \quad s \leq s_M = 5 \text{ cm} \\ 0 & \text{if} \quad s > s_M = 5 \text{ cm} \end{cases}$$

We then calculated the following:

1. If we choose to transplant every patient with any tumor size equal or less than a critical tumor size, S_F, then to guarantee that all patients with such tumor size less than S_F are transplanted (that is, all grafts are used), S_F has to be defined as

$$N\int_0^{s_F} p(s)ds = F \tag{10.11}$$

or

$$s_F = -\frac{\log\left(1 - \frac{F}{N}\right)}{\lambda} \tag{10.12}$$

In other words, this means to choose a policy such that $x(s \leq s_F) = p(s)$ and $y(s \leq s_F) = 0$.

Eqs. (10.11) and (10.12) can be interpreted as follows: $\int_0^{S_F} p(s)ds$ is the fraction of the population that has tumors of size equal or less than S_F. Multiplied by the total population N gives the total number of individuals that are transplanted, that is, received all the liver grafts F. In other words, all available livers are used. The size limit that guarantees that this happens, S_F, is therefore calculated as a function of F as in Eq. (10.12).

2. Hence, if not all patients with tumor size s are transplanted, for example, if we choose to transplant $x(s) = \frac{F}{N}p(s)$ and not transplant $y(s) = \left(1 - \frac{F}{N}\right)p(s)$, then we can choose to transplant all the patients with tumor size up to $s_0 > s_F$.
3. Using MC (see above), the proportion of nontransplanted patients with tumor size s below S_M with respect to the total number of HCC patients at presentation is

$$p_{nt}(s < s_M) = \begin{cases} \dfrac{N(1 - e^{-\lambda s_M}) - F}{N} & \text{if } F < N(1 - e^{-\lambda s_M}) \\ 0 & \text{otherwise} \end{cases} \quad (10.13)$$

Eq. (10.13) means that multiplying the proportion of patients with tumor size equal or less than S_M, $(1 - e^{-\lambda S_M})$ by the total population of HCC, N, gives the number of patients with tumors of size up to S_M. This number minus the number of available livers divided by the total population size gives the proportion of nontransplanted patients.

4. The proportion of transplanted patients with respect to the total number of HCC patients at presentation, with tumor size s below S_M, is

$$p_t(s < s_M) = \begin{cases} \dfrac{F}{N} & \text{if } F < N(1 - e^{-\lambda s_M}) \\ \dfrac{F}{N}(1 - e^{-\lambda s_M}) & \text{otherwise} \end{cases} \quad (10.14)$$

Note that in the exceptional and unique case when $s_F = s_M$ all the grafts are used (see Models' Limitations for a more thorough discussion).

Eq. (10.14) reflects the fact that a fraction $\frac{F}{N}$ of those individuals with tumor size equal or less than s_M is transplanted when the number of available livers F is less than the number of individuals with tumor size greater than s_M at presentation.

5. If the MC are obeyed, then the proportion of nontransplanted patients with tumor size greater than s_M is

$$p_{nt}(s > s_M) = e^{-\lambda s_M} \quad (10.15)$$

which is the minimum (if F is not enough to transplant up to s_M) proportion of individuals with tumor size greater than s_M. According to MC, none of those patients are transplanted, independently of F.

6. If MC are not obeyed, then there is a proportion of transplanted patients with tumor size greater than s_M that could be transplanted. This proportion is limited by the number of available livers, and it is

$$p_t(s > s_M) = \begin{cases} \dfrac{Fe^{-\lambda s_M}}{N} & \text{if } F > N(1 - e^{-\lambda s_M}) \\ 0 & \text{otherwise} \end{cases} \tag{10.16}$$

In this situation, the proportion of nontransplanted is given by

$$p_t(s > s_M) = \begin{cases} e^{-\lambda s_M} & \text{if } F < N(1 - e^{-\lambda s_M}) \\ 1 - \dfrac{F}{N} & \text{otherwise} \end{cases} \tag{10.17}$$

Note that adding the proportion of nontransplanted individuals with tumor sizes greater and less than s_M gives $1 - \frac{F}{N}$. By the same token, adding the proportion of transplanted individuals with tumor sizes greater and less than s_M gives $\frac{F}{N}$.

7. Now, we abandon the MC and transplant a proportion $x(s) = \frac{F}{N}p(s)$ of individuals with tumor size up to s_0 (variable) and compare the impact on the total mortality of HCC patients with the mortality resulting from adopting the MC.

First, we calculate the survival of transplanted patients (*TS*) with tumor size up to s_0 at a moment in time T after the patient's presentation. The proportion of the individuals with tumor size up to s_0 at presentation is

$$\int_0^{s_0} \lambda e^{-\lambda s} ds \tag{10.18}$$

The proportion of patients at presentation who were transplanted and survived up to T after the transplantation is

$$\int_0^{s_0} x(s)e^{-\alpha_t(s)T} ds = \frac{F}{N} \int_0^{s_0} \lambda e^{-\lambda s} e^{-\alpha_t(s)T} ds \tag{10.19}$$

Hence, the total number of transplanted patients (TS) with tumor size up to S_0 at presentation and who survived up to time T is given by Eq. (10.19) multiplied by N

$$TS = N\frac{F}{N}\int_0^{s_0} \lambda e^{-\lambda s} e^{-\alpha_t(s)T} ds = F\int_0^{s_0} \lambda e^{-\lambda s} e^{-\alpha_t(s)T} ds \tag{10.20}$$

8. The number of patients, who were not transplanted, with tumors up to tumor size S_0 at presentation is

$$N\int_0^{s_0} y(s)ds = N\int_0^{s_0}\left(1-\frac{F}{N}\right)\lambda e^{-\lambda s} ds \tag{10.21}$$

and those who survived after time T are

$$N\int_0^{s_0}\left(1-\frac{F}{N}\right)\lambda e^{-\lambda s} e^{-\alpha_{nt}(s)T} ds \tag{10.22}$$

Now, the number of patients with tumors greater than size s_0 at presentation, which were not transplanted, is

$$N\int_{s_0}^{\infty} p(s)ds = N\int_{s_0}^{\infty} \lambda e^{-\lambda s} ds \tag{10.23}$$

and, among those, the survivors after time T are

$$N\int_{s_0}^{\infty} \lambda e^{-\lambda s} e^{-\alpha_{nt}(s)T} ds \tag{10.24}$$

Hence, the total number of survivors after time T who were not transplanted is

$$NTS = N\int_0^{s_0}\left(1-\frac{F}{N}\right)\lambda e^{-\lambda s} e^{-\alpha_{nt}(s)T} ds + N\int_{s_0}^{\infty} \lambda e^{-\lambda s} e^{-\alpha_{nt}(s)T} ds \tag{10.25}$$

9. Therefore, the total survival is obtained by adding Eqs. (10.20) and (10.25):

$$\begin{aligned}\text{Survivors} = {} & F\int_0^{s_0} \lambda e^{-\lambda s} e^{-\alpha_t(s)T} ds + N\int_0^{s_0}\left(1-\frac{F}{N}\right)\lambda e^{-\lambda s} e^{-\alpha_{nt}(s)T} ds \\ & + N\int_{s_0}^{\infty} \lambda e^{-\lambda s} e^{-\alpha_{nt}(s)T} ds\end{aligned} \tag{10.26}$$

10. Finally, the total mortality is given by

$$M(s_0) = N - \left[F \int_0^{s_0} \lambda e^{-\lambda s} e^{-\alpha_t(s)T} ds + N \int_0^{s_0} \left(1 - \frac{F}{N}\right) \lambda e^{-\lambda s} e^{-\alpha_{nt}(s)T} ds + N \int_{s_0}^{\infty} \lambda e^{-\lambda s} e^{-\alpha_{nt}(s)T} ds \right] \quad (10.27)$$

11. Now, to calculate the optimal transplantation strategy, we determine the tumor size that can be transplanted and find either s such that $\min[M(s)]$ or s such that $M(s) = M(s_M)$.

Results

We illustrate the above analysis for a simulation of a theoretical population of 1500 HCC patients with tumor size parameter distribution of λ equal to 0.3. As the total number of patients in the real samples from which data were retrieved was 327 patients, this implied in an average size of 3.3 cm and a 95% confidence interval of [2.9; 3.7] (Yao et al., 2001a,b; Bismuth et al., 1993; Marsh et al., 2000). The total number of available livers to be grafted was assumed to be 500. With this, we simulated the total number of deaths in both transplanted and nontransplanted HCC patients after 5 years as a function of the tumor size of transplanted patients. The result is shown in Fig. 10.3.

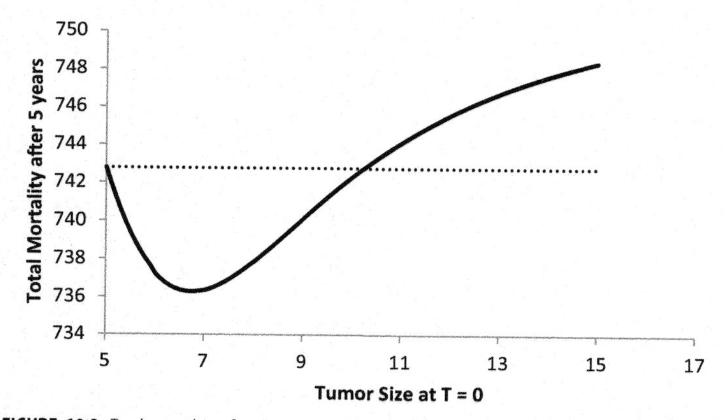

■ **FIGURE 10.3** Total mortality after 5 years comprising both transplanted and nontransplanted HCC patients in a 1500 theoretical population. We show only what happens when individuals have tumor size greater than the strict MC (5 cm).

Fig. 10.3 shows the total mortality in the HCC patient cohort, including those transplanted and those nontransplanted as well. The dotted line is a reference line: the point where the mortality curve crosses it is the maximum tumor size that could be transplanted without worsening the mortality in the list. Note that it is possible to include patients with tumor size up to 10 cm without increasing the total mortality of this cohort.

Conclusions

Some limitations of the model should be highlighted. Firstly, the most important is the fact that we considered a cohort of HCC patient isolated from the others causes of liver failures and, therefore, from the waiting list. We circumvent this by assuming that the 500 available grafts were the equivalent of the number of livers typically allocated to this kind of patients. Secondly, we arbitrarily assumed an exponential distribution for the tumor size, although this is likely to be true. Thirdly, we assumed an ad hoc function for the death rates of transplanted and nontransplanted patients. However, assuming any convex function for transplanted mortality rate as a function of tumor size and concave function for nontransplanted mortality rate would not qualitatively modify our results. Finally, on important consequence of the model, although not directly observable from the equations, is that by transplanting patients with tumor size greater than S_F, and, therefore, not transplanting a proportion of patients with tumor size less than S_F may result in a certain proportion of F livers that will not be used. This is a consequence of Eq. (10.20) when $T=0$, that is, $F\int_0^{s_0} \lambda e^{-\lambda s} ds = F\left(1 - e^{-\lambda s_0}\right)$. Note, however, that this would happen with any model that would not transplant all the patients below a certain tumor size when there are enough livers available. Eq. (10.14) illustrates that if there are enough livers then everybody with tumor size below S_M would be transplanted. This should not be taken as an advantage of MC because (as can be seen from Eq. 10.14) transplanting every patient in need is an exceptional case that occurs when $s_M = s_F$. Remember that s_M is determined by law and s_F by chance, depending on the number of available grafts F, the incidence of HCC, N, and the distribution of tumor size in these patients at presentation.

The MELD (Model for End-Stage Liver Disease) score has been selected as the most clinically appropriate tool for accurately predicting mortality in patients with chronic liver diseases (Chaib and Massad, 2005, 2007; 2008a,b,c). However, the MELD score does not accurately predict survival in some patients, such as those with HCC. To enable patient with HCC to undergo LT at a rate similar to that for patients without HCC, additional points based on the number and size of the HCC nodules are assigned to patients with HCC on the waiting list; the intention is to match the risk of death for those with similar MELD scores but no HCC (Chawla et al., 2011). With this strategy, HCC patients have easier access to transplantation than non-HCC ones. In addition, this system does not allow for a dynamic assessment, which would be required to picture the current use of local tumor treatment.

Because of the paucity of donor organs, efforts have been made to optimize the effectiveness of LT through the application of strict criteria for selecting patients who have the greatest likelihood of prolonged survival after surgery. LT is a well-established treatment in a subset of patients with cirrhosis and HCC. MC (single nodule up to 5 cm, up to three nodules none larger than 3 cm, with no evidence of extrahepatic spread or macrovascular invasion) have been traditionally accepted as standard of care. The introduction of MC improved 5 year survival post-LT for HCC from below 50% to greater than 70% (Ringe et al., 1991; Iwatsuki et al., 1991a,b). However, some groups have proposed that these criteria are too restrictive and exclude some patients from transplantation who might benefit from this procedure. Transplanting patients with tumors beyond the established criteria falls into two categories, those whose tumors are beyond the MC at presentation without the use of treatment prior to transplantation (expanded criteria) and those in whom treatment allows the MC to be fulfilled (downstaging). Currently, however, there is no international consensus regarding these approaches in clinical practice, as different populations such as Europeans, Americans, or Asians have distinct HCC evolution, and this should greatly influence the establishment of transplantation criteria (Campos Freire et al., 1968; Machado, 1972).

Expanded Milan criteria (EMC) can be defined by the use of LT in recipients with tumors beyond the MC. The first description was published in 2001 by the group of the University of California, San Francisco (UCSF) (Figueras et al., 2000). In their study, 70 HCC patients who underwent LT were retrospectively evaluated on the basis of explant analysis, not pretransplant radiology. In the 60 cases with either a single nodule up to 6.5 cm, or up to three nodules none larger than 4.5 cm, and total tumor diameter no more than 8 cm, the 5-year overall survival was 75.2%. 46 out of the 60 patients (76%) had tumors that were within the MC, and these had a 5-year survival of 72%. Subsequently, a number of different EMC proposals have been described (Onaca et al., 2007; Roayaie et al., 2002).

To optimize allocation of donated organs, Volk et al. (2008) created a mathematical model focused on the lowest acceptable survival rate after LT for which the use of donor organs of standard quality could be justified and revealed that unless a 5-year survival of at least 61% could be achieved, performing LT for patients with tumors beyond MC put other patients without HCC at a risk of dying without LT (Volk et al., 2008). This survival rate may increase to 71% in regions with severe organ shortage and reduced 25% in regions where shortage is not so acute. Samuel et al. (2011) comment that this study has several limitations because it did not evaluate the use of donor organs of marginal quality, and it assumed that long-term survival after LT does not vary as a function of preoperative MELD score.

More recently, Tosa et al. (2012) used a competitive risk model assessment and suggest a model for comparing the opportunities of receiving a graft for both HCC (deMELD) and non-HCC (MELD) patients on a common waiting list concluding that the allocation of deMELD (dropout risk scores to HCC) has the potential to allow for a dynamic and combined comparison of opportunities to receive a graft for HCC and non-HCC patients on a common waiting list.

There is a lack of studies addressing these issues in the literature. In addition, the extrapolation of these findings to routine clinical practice is limited by our inability to accurately predict survival after LT for individual patients with HCC who do not meet the MC.

Finally, the methodology used in this chapter explored the theoretical outcomes of HCC patients as a function of tumor size for transplantation, violating the limit proposed by the Milan criterion. The model proposed was based on the calculation of mortality as a function of tumor size. Other indicators of clinical outcomes could be used instead of tumor size with the same model. In addition, other methods of analysis could be used to optimize the number of patients who could be transplanted, such as game theory (Massad et al., 2005) or nonbinary logics such as the theory of fuzzy sets (Massad et al., 1999, 2008a,b; Ortega et al., 2003). This, however, will be subject of future work.

10.2 LIVER TRANSPLANTATION AND EXPANDED MILAN CRITERIA: DOES IT REALLY WORK?

Introduction

Orthotopic liver transplantation (OLT) is an excellent approach for HCC in well-selected candidates (Busuttil et al., 2005; Li et al., 2009). The most widely used criteria for patient selection are those proposed by Mazzaferro et al. (1996), the so-called MC (a single tumor up to 5 cm or up to 3 tumors none larger than 3.0 cm).

When MC were applied, there was a significant improvement in survival over time for HCC patients undergoing OLT with a 5-year survival of 61, 1% contrasting with previously observed 5-year survival rate of 25, 3% in 1987.

Methods

A systematic search of Medline (PubMED) database was performed to identify studies evaluating expanded criteria for patients with HCC submitted to liver transplantation. The search was restricted to papers written in English and published from 2000 to 2009. The keywords used were HCC, liver transplantation, expanded criteria, the UCSF criteria, MC, and others. Only papers reporting cadaveric liver donors that evaluated expanded criteria on the basis of tumor number and size were selected. This search resulted in a total of 39 studies. Additionally, a full manual search from bibliographies of papers describing aspects beyond tumor number and size and reports of consensus conference was also performed. Nineteen papers were excluded due to irrelevance of subject, lack of information, and incompatibility of language.

We compiled data focusing on patient survival rate and tumor recurrence-free rate from 1 to 5 years. We also compared results between MC and EMC. $P < 0.05$ was considered statistic significant.

In recent years, however, some groups have argued that the MC are too restrictive and exclude some HCC patients from OLT despite the possibility of benefit. Expanded criteria can be defined by the use of OLT in recipients with tumors beyond the MC. The first description was published in 2001 by the group of the University of California. In their study, 70 HCC patients who underwent OLT were retrospectively evaluated on the basis of explants analysis, not pretransplant radiology. In the 60 cases with either a single nodule up to 6.5 cm, or up to three nodules none larger than 4.5 cm, and total tumor diameter no more than 8 cm, the 5-years overall survival was 75.2%. 46 out of the 60 patients (76%) had tumors that were within the MC, and these had a 5-year survival of 72%.

Subsequently, in the past 10 years, some major new criteria were created expanding MC (EMC) such as UCSF criteria, Kyoto (Takada et al., 2007), Asian (Lee et al., 2008), Shanghai (Fan et al., 2009), and others.

Our aim is to study the current situation of these several EMC proposed to clarify both this debate through a critical analysis of available data and addressed discussion of further parameters beyond number and size of tumors, focusing on patient survival rate and tumor recurrence-free rate from 1 to 5 years after OLT.

Results

Twenty-three papers including centers from North and South America, Europe, and Asia were compiled. Fourteen different EMC were found; however, UCSF was the most studied (Decaens et al., 2012; Duffy et al., 2007; Takada et al., 2007; Fan et al., 2009; Lee et al., 2008). The patient survival rate and tumor recurrence-free rates from 1 to 5 years are shown in the table below (Table 10.1).

Table 10.1 Comparison of patients' survival rate and tumor recurrence-free rate among different expanded Milan criteria in the literature.

Author	Year	N	Criteria	1y	3y	5y	Staging
Mazzaferro et al.	2009	283	Up-to-seven: in the absence of intravascular invasion, fulfilled the so-called up-to-seven criteria, with seven being the result of the sum size (in cm) and number of tumors for any given hepatocellular carcinoma.	–	77.7%	72.1%	P
Fan et al.	2009	176	Single tumor ≤9 cm in diameter, 2–3 tumors with the largest ≤5 cm, a total tumor diameter ≤9 cm without macrovascular invasion, lymph node invasion, and extrahepatic metastasis.	82.7%/ 51.2%[b]	–	79.9%/ 46.1%[b]	P
Li et al.	2009	25	Total tumor size ≤9 cm which were without macrovascular invasion or extrahepatic metastases, regardless of the number of tumor lesions	85.2%/ 95.0%[b]	–	48.4%/ 70.4%[b]	P

Table 10.1 Comparison of patients' survival rate and tumor recurrence-free rate among different expanded Milan criteria in the literature. *continued*

Author	Year	N	Criteria	1y	3y	5y	Staging
Takada et al.	2010	23	Diameter ≤5 cm; number of lesions ≤10; PIKVA-II ≤ 400 mAU/mL.	–	–	95%[b]	R
Boin et al.	2008	19	Any tumor beyond MC.	–	–	47.4% (10y)	P
Lee et al.	2008	22	Largest tumor ≤5 cm; HCC number ≤6; no gross vascular invasion.	100%	88.9%	80%	P
Majeed et al.	2008	23	No extrahepatic evidence of spread on abdomen computed tomography and chest.	–	50%/70%	–	NA
Silva et al.	2008	26/46[a]	Up to 3 nodules, with none larger than 5 cm, and a cumulative tumor burden ≤10 cm.	92%/74%[a]	79%/55%[a]	69%/40%[a]	R/P
Herrero et al.	2008	24	Any tumor beyond Milan Criteria	92%	78%	73%	R
Onaca et al.	2007	130	2–4 tumors with the largest ≤5 cm or a single tumor up to 6 cm.	–	–	64.6%[b]/NA	P
Soejima et al.	2007	40	Without extrahepatic spread or macroscopic vascular invasion. The size and number of HCC nodules were not limited.	83%[b]	74%[b]	–	R
Todo et al.	2007	272	Any tumor beyond the Milan Criteria.	–	–	66.4%[b]	P
Takada et al.	2007	33	≤10 tumors and all ≤5 cm.	–	–	93%[b]	R
Hwang et al.	2005	62	Any tumor beyond the Milan Criteria.	–	62.6%	–	P
Knetman et al.	2004	21	1 nodule <7.5 cm any number <5.	90.5%	–	83% (4y)	R
Todo et al.	2007	171	Any tumor beyond the Milan Criteria.	75%/64.9%[b]	60.4%/52.6%[b]	–	NA
Roayaie et al.	2002	32	1 or more nodules 5–7 cm.	–	–	55%[b]	R

N = *number of patients transplanted in each study. Staging refers to the method used for tumor staging*— NA, *not available;* P, *pretansplant radiology;* R, *explant tumor pathology.*
[a]*Pathology based.*
[b]*Recurrence-free rate.*

Conclusions

HCC is a major health problem worldwide (Llovet et al., 2003). In the west, 30%–40% of HCC cases are detected at early stages and treated with intention to cure, a figure that reaches 60% of the cases in Japan (Llovet et al., 2003). Surgical treatments are accepted as the standard of care for early tumors because they provide survival rates consistently better than their untreated counterparts (5-years survival rates of 40%–70% vs. < 20%) (Llovet et al., 2003, 2005; Bruix et al., 2005). Resection of single tumors in patients with well-preserved liver function leads to remarkable outcomes (5-years survival exceeds 50%–60%) (Llovet et al., 2005). Early results after OLT in unselected patients with cirrhosis and HCC were poor, with early recurrence rates and 5-year survival of only 18%–49% (Iwatsuki et al., 1991a,b; Ringe et al., 1991; Neuhaus et al., 1999; Llovet et al., 2000; Roayaie et al., 2004). Several small studies in the early 1990s suggested that recurrence-free survival could be improved by restricting transplantation to patients with two–three nodules or a single tumor < 3–5 cm in diameter (Bismuth et al., 1993; Figueras et al., 1997; Sauer et al., 2005). Two large retrospective studies (Iwatsuki et al., 1991a,b; Klintmalm et al., 1998) confirmed that tumors > 5 cm had a high rate of post-transplantation recurrence, largely because of the association with vascular invasion and poor differentiation.

MC are considered the gold standard for selection of the best HCC candidates for OLT after numerous external validations of the seminal proposed (Mazzaferro et al., 1996). In fact, the MC as a restriction selection for patients with HCC have been confirmed as consistent by several other groups among more than 1000 patients (Bismuth et al., 1999; Jonas et al., 2001; Llovet et al., 1998). The MC were subsequently used by the United Network for Organ Sharing to assign the listing priority of patients presenting HCC.

On the other hand, some studies have recently suggested that the MC might be too restrictive, with relatively good results achieved when different proposals are used (Table 10.2).

There are essential aspects that should be considered when treatments related to HCC are evaluated: (1) treatments that achieve survival rates higher than 50% in 5 years are considered effective therapies, given the fact that studies have demonstrated the 3-years survival of early HCC to be about 50% (Llovet et al., 1999a,b; Bruix et al., 2003); (2) the deleterious impact of the progressive increase in the waiting list time has to be considered when the efficacy of OLT as a treatment for HCC is evaluated because of the risk of tumor progression and death during this period (Llovet et al.,

Table 10.2 Comparison of patient survival rate and tumor recurrence-free rate between University of California San Francisco (UCSF) and Milan criteria.

Author (year)	n		Staging	3 year				5 year			
				USFC		Milan		UFSC		Milan	
	USFC	Milan		PS	RR	PS	RR	PS	RR	PS	RR
Yao et al. (2001a,b)	14	46	–	–	–	–	–	73%	–	72%	–
Decaens et al. (2012)	44	279	R(0.10)[ITT]/ (0.14)[a]	–	–	–	–	45.6%[ITT]	47.80%	60.10%[ITT]	60.40%
Yao et al. (2007)	39	184	P(0.33)¥/ (0.26)[a]	–	–	–	–	63.60%	62.70%	70.40%	70.20%
	38	130	R(0.58)[a]	–	–	–	–	82%	93.60%	80%	90.10%
Duffy et al. (2007)	185	173	R(0.061)	74%	–	–	–	64%	–	79%	–
	208	126	P(0.057)	83%	–	–	–	71%	–	86%	–
Lee et al. (2008)	10	152	P(0.953)	90%	–	–	–	78.8%	–	86%	–
Xiao et al. (2009)	32	68	P(0.058)/ (0.103)[a]	55.2%	40%	88.4%	81.8%	–	–	–	–

ITT = intention to treat; ¥ = patient survival (PS). Staging refers to the method used for tumor staging— NA, *not available; P, pretansplant radiology; R, explant tumor pathology.*
[a]*Recurrence-free rate (RR).*

1999a,b); and (3) it is well known that preoperative imaging techniques underestimated HCC staging in about 20% of cases, and thus, the extrapolation of the histopathologic data to the preoperative scenario might be misleading (Burrel et al., 2003).

Yao et al. (2001a,b) from UCSF reported a 5-year survival of 75% in patients with single tumor as large as 6.5 cm or a maximum of three tumors up to 4.5 cm and a cumulative tumor burden <8 cm. With mostly retrospective data, some groups have independently tested these criteria (Decaens et al., 2012; Duffy et al., 2007). These results have, however, been challenged because of the use of explants pathology, rather than preoperative imaging, as a determinant for the definition of the tumor stage.

The UCSF proposal is the approach mostly tested; however, it has been challenged because of the use of explants pathology. Duffy et al. (2007) and Yao et al. (2001a,b) recently published their results analyzing the survival rates and recurrence probabilities on the basis of the pre-OLT radiologic assessment.

Expansion of tumor criteria for transplantation risks includes patients with high-grade tumors or microvascular invasion who might have a higher risk of recurrence. Indeed, as observed in previous reports, we found that tumors that exceeded the MC criteria were more likely to have evidence of vascular invasion in the explants (Iwatsuki et al., 1991a,b; Klintmalm et al., 1998). However, interestingly, this finding only affected recurrence-free survival when there were > five tumors or a single tumor >6 cm. Thus, this may reflect tumor mass and the degree of vascular invasion rather than the presence of vascular invasion itself.

Several studies have shown that some tumor patients transplanted outside of the UNOS and MC survive longer (Cillo et al., 2004; Durand and Belguiti, 2003; Gondolesi et al., 2002, 2004; Herrero et al., 2001; Ringe et al., 1991; Scwartz, 2004), and as consequence, several proposals have been made to expand the HCC inclusion criteria (Herrero et al., 2001; Llovet et al., 2004; Yao et al., 2001a,b). However, the criteria proposed by the University of Pittsburgh and some groups in Europe are based, at least in part, on pathological features (nodal invasion, grade, vascular invasion) that are not usually available before transplantation (Iwatsuki et al., 1991a,b; Cillo et al., 2004; Santoyo et al., 2005; Zavaglia et al., 2005). By contrast, MC, UCSF, and Onaca et al. (2007) proposal rely on factors (tumor size and number) that can be determined by preoperative imaging; however, such criteria must consider the limitations of imaging studies (Rode et al., 2001; Stoker et al., 2002; Burrel et al., 2003; Steingruber et al., 2003; Teefy et al., 2003; Kim et al., 2004). The continued improvement in imaging techniques may decrease the gap between imaging and pathology of HCC, although some understanding will certainly continue to exist (Silva et al., 2008).

Silva et al. (2008) reported 281 cases of HCC in cirrhotic livers treated by OLT using a new criteria (up to three tumors, each no larger than 5 cm, and a cumulative tumor burden < 10 cm) with a 5-year survival rate of 57% based on the intention-to-treat principle. The 5-year survival rate was 63% among transplanted patients.

No difference in survival or in recurrence was found between cases within and beyond the UCSF criteria and others. Although these results suggest that this expansion does not result in an impaired outcome, we understand that they need validation, given the relatively small number of patients. Microvascular invasion was the only factor that predicted poor survival in the multivariate analysis. Indeed, several studies have shown that the differentiation degree and microvascular invasion represent direct indicators of the biologic progression of HCC, being associated with tumor recurrence and poor long-term survival (Herrero et al., 2001; Moya et al., 2002; Roayaie et al., 2002; Cillo et al., 2004; Duffy et al., 2007).

Expansion of transplantation inclusion criteria should be made cautiously. Listing criteria for HCC should reflect the minimum acceptable recurrence-free survival rate and must reflect a consensus of transplantation community. Furthermore, the results need to be confirmed prospectively if criteria were liberalized to ensure that an unrecognized selection bias did not influence the results themselves. Finally, the societal benefit of expanding tumor criteria needs to be weighed against a relatively fixed donor organ supply and a growing demand for OLT for other indications, such as decompensated cirrhosis due to chronic hepatitis C, where long-term survival may be better.

In conclusion, the EMC are a useful attempt for widening the preexistent protocol for patients with HCC in the waiting list for OLT; however, there is no significant difference in patient survival rate and tumor recurrence-free rate from those patients who followed the MC.

Theoretical impact of HBV and HCV vaccines on liver transplantation

11.1 THEORETICAL IMPACT OF AN ANTI-HCV VACCINE ON THE ANNUAL NUMBER OF LIVER TRANSPLANTATION

Introduction

The development of an effective vaccine for hepatitis C is of paramount importance, given the global disease burden and its public health impact (Stoll-Keller et al., 2009).

The hepatitis C virus (HCV) infects some 170 million people worldwide and is responsible for approximately 20% of cases of acute hepatitis and 70% of cases of chronic hepatitis.

Currently in our country (Brazil), there are, approximately, 3.5 million people infected with HCV virus, needless to say that in our waiting list for liver transplantation 50% of the cases are HCV-related end-stage cirrhosis.

Treatment with the combination of pegylated interferon and ribavirin induces a sustained virological response in roughly 40% of the patients; therefore, we are far from reach of the best treatment for this disease.

In addition, HCV-related end-stage liver diseases and its consequence such as hepatocellular carcinoma (HCC) have become the leading cause for liver transplantation worldwide (Tanwar et al., 2009; Gallegos-Orozco et al., 2009).

Mathematical Approaches to Liver Transplantation. https://doi.org/10.1016/B978-0-12-817436-4.00011-4

Our aim is to analyze, through a mathematical model, the potential impact of a theoretical anti-HCV vaccine on the number of liver transplantations in the years ahead.

Hypothesis

We hypothesize that the advent of an effective vaccine against the HCV will take several years to have a significant impact on the number of liver transplantation.

The model

The model was originally proposed to study hepatitis C (Massad et al., 2009), without treatment or vaccination, and it assumes a population divided into 14 states. Susceptible individuals, denoted $S(t)$, are subdivided into two classes, denoted as $S_1(t)$ representing the general population and $S_2(t)$ representing a group with higher risk of acquiring the infection by HCV, such as injecting drug users, recipients of blood/blood products transfusion, people occupationally exposed to blood/blood products, etc. $S_1(t)$ individuals are transferred to the high-risk group $S_2(t)$ with rate ξ. Both susceptible classes acquire the infection with rates β_I ($I = 1, \ldots, 4$) from four different types of individuals, namely, infected and asymptomatic but already infectious individuals, denoted $HCV(t)$, individuals with acute hepatitis, denoted $Ac(t)$, and individuals with intermediate fibrosis stages, called "chronic state," denoted $Chr(t)$. Only a fraction u of β_I is effective for people in the nonrisk group. Susceptible individuals can develop other hepatopathies causing liver failure with rate ωω, after which they are transferred to the state called $Oth(t)$. Infected and asymptomatic individuals, $HCV(t)$, can evolve to acute or chronic hepatitis, with rates σ_1 and σ_2, respectively. These individuals evolve to $Chr(t)$ at rate δ_1. Once in the chronic state, individuals can either evolve to HCC, denoted $HCC(t)$, at rate θ, or evolve to liver failure at rate δ_2. We consider only four levels of the Model for End-Stage Liver Disease (MELD), a scale of hepatopathy that incorporates three widely available laboratory variables including the international normalized ratio (INR), serum creatinine, and serum bilirubin. Those individuals are denoted as $MELD_I$ ($I = 1, \ldots, 4$). Evolution between the MELD levels occurs at rates ε_1 ($I = 1, \ldots, 3$). Infected individuals without chronic hepatitis can recover to compartment $R(t)$ at rates γ_I ($I = 1, \ldots, 4$). Individuals from compartments $MELD_I$ ($I = 1, \ldots, 4$) and $HCC(t)$ get into the waiting list for liver transplantation, $WL(t)$, at rates φ_I ($I = 1, \ldots, 6$). Individuals in the waiting list are eventually transplanted at rate τ_1.

We also consider the possibility of lost of graft at rate φ_6 and a secondary waiting list $WLTx(t)$. From the latter, individuals are eventually transplanted at rate τ_2. Every individual in this population is subjected to a mortality rate μ and an additional mortality of those with liver disease occurs at rates α_I ($I = 1, \ldots, 10$) depending on the state. Finally, the susceptible compartment grows at rate Λ, which is comprised by the sum of all mortalities to keep the total population constant.

The model's structure is summarized in Fig. 11.1.

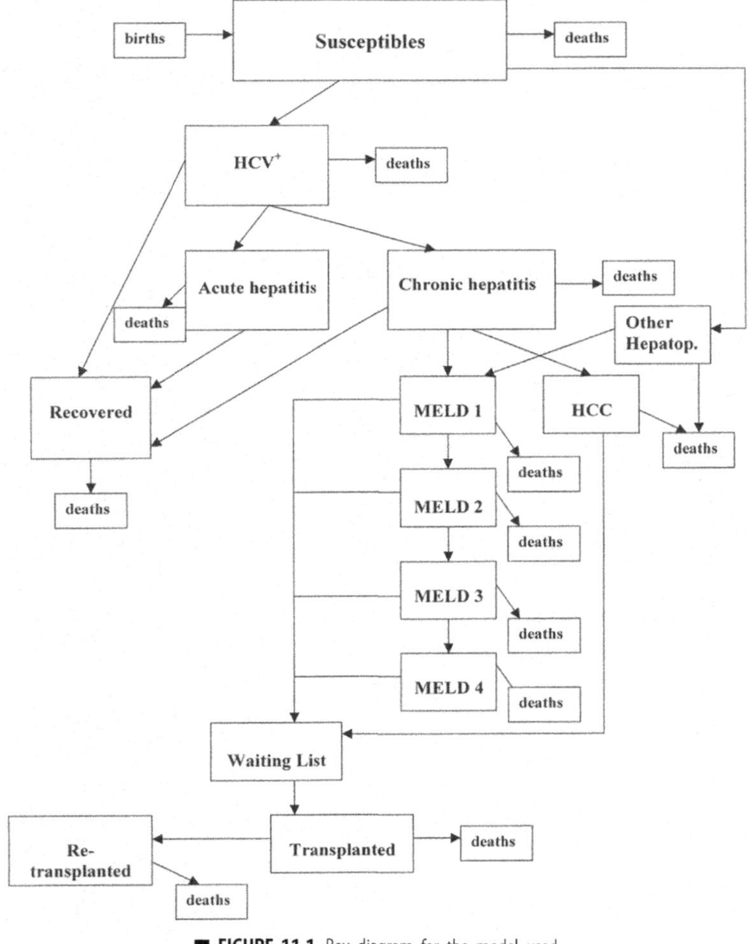

■ **FIGURE 11.1** Box diagram for the model used.

The model is described by the following set of differential equations:

$$
\begin{aligned}
\frac{dS_1(t)}{dt} &= \Lambda - u\beta_1 HCV(t)\frac{S_1(t)}{N} - u\beta_2 Ac(t)\frac{S_1(t)}{N} - u\beta_3 T_A(t)\frac{S_1(t)}{N} \\
&\quad -u\beta_4 Chr(t)\frac{S_1(t)}{N} - (\mu + \omega + \xi)S_1(t) \\
\frac{dS_2(t)}{N} &= \xi S_1(t) - \beta_1 HCV(t)\frac{S_2(t)}{N} - \beta_2 Ac(t)\frac{S_2(t)}{N} - \beta_3 T_A(t)\frac{S_2(t)}{N} \\
&\quad -\beta_4 Chr(t)\frac{S_2(t)}{N} - (\mu + \omega)S_2(t) \\
\frac{dHCV(t)}{dt} &= \beta_1 HCV(t)\frac{(S_1(t) + S_2(t))}{N} + \beta_2 Ac(t)\frac{(S_1(t) + S_2(t))}{N} \\
&\quad + \beta_3 T_A(t)\frac{(S_1(t) + S_2(t))}{N} + \beta_4 Chr(t)\frac{(S_1(t) + S_2(t))}{N} \\
&\quad -(\gamma_1 + \sigma_1 + \sigma_2 + \mu)HCV(t) \\
\frac{dAc(t)}{dt} &= \sigma_1 HCV(t) - (\delta_1 + \gamma_2 + \alpha_1 + \mu)Ac(t) \\
\frac{dR(t)}{dt} &= \gamma_1 HCV(t) + \gamma_2 Ac(t) + \gamma_4 Chr(t) - \mu R(t) \\
\frac{dChr(t)}{dt} &= \delta_1 Ac(t) + \sigma_2 HCV(t) - (\delta_2 + \theta + \alpha_2 + \gamma_3)Chr(t) \\
\frac{dMELD_1(t)}{dt} &= \delta_2(Chr(t) + Oth(t)) - (\varphi_1 + \varepsilon_1 + \alpha_3 + \mu)NT_C M_1(t) \\
\frac{dMELD_2(t)}{dt} &= \varepsilon_1 NT_C M_1(t) - (\varphi_2 + \varepsilon_2 + \alpha_4 + \mu)NT_C M_2(t) \\
\frac{dMELD_3(t)}{dt} &= \varepsilon_2 NT_C M_2(t) - (\varphi_3 + \varepsilon_3 + \alpha_5 + \mu)NT_C M_3(t) \\
\frac{dMELD_4(t)}{dt} &= \varepsilon_3 NT_C M_3(t) - (\varphi_4 + \alpha_6 + \mu)NT_C M_4(t) \\
\frac{dHCC(t)}{dt} &= \theta Chr(t) - (\varphi_5 + \alpha_7 + \mu)HCC(t) \\
\frac{dWL(t)}{dt} &= \sum_{i,j=1}^{4} \varphi_i NT_C M_j(t) + \varphi_5 Chr(t) - (\tau_1 + \alpha_8 + \mu)WL(t) \\
\frac{dTx_1(t)}{dt} &= \tau_1 WL(t) - (\varphi_6 + \alpha_9 + \mu)Tx_1(t) \\
\frac{dWLTx(t)}{dt} &= \varphi_6 Tx_1(t) - (\tau_2 + \alpha_{10} + \mu)WLTx(t) \\
\frac{dTx_2(t)}{dt} &= \tau_2 WLTx(t) - (\alpha_{11} + \mu)Tx_2(t) \\
\frac{dOth(t)}{dt} &= \omega(S_1(t) + S_2(t)) - (\delta_2 + \alpha_{12} + \mu)Oth(t)
\end{aligned}
\tag{11.1}
$$

Table 11.1 Parameters used in the simulations.

Parameters	Biological meaning	Numerical values
Λ	Births	Variable
u	Attenuation of β_i affecting people in the nonrisk group	0.05
ξ	Rate of transfer to high-risk group	1.0×10^{-5} days^{-1}
μ	Natural mortality rate	3.9×10^{-5} days^{-1}
β_i	Potentially infective contact rates	3.0×10^{-3} days^{-1}
γ_i	Recovery rates from infection	From 5.0×10^{-5} to 1.0×10^{-4} days^{-1}
σ_i	Rates of evolution to hepatitis	From 1.0×10^{-3} to 9.0×10^{-3} days^{-1}
δ_1	Rates of evolution from acute to chronic hepatitis	7.0×10^{-5} days^{-1}
δ_2	Rate of evolution to hepatic failure (MELD)	5.0×10^{-5} days^{-1}
θ	Rate of evolution of hepatocellular carcinoma	1.0×10^{-6} days^{-1}
φ_i	Rates of getting into the waiting lists	From 1.1×10^{-7} to 2.0×10^{-6} days^{-1}
τ_i	Transplantation rates	Variable
ω	Rate of evolution to other hepatopathies	1.3×10^{-6} days^{-1}
α_i	Disease-induced mortality rates	From 4.2×10^{-5} to 5.9×10^{-5} days^{-1}

The biological meaning and the values of the parameters can be seen in Table 11.1.

Testing the hypothesis

Model's simulation

The model's equations are described with detail in Chaib and Massad (2008a,b,c) and it was simulated with the same set of parameters as in Nicholson et al. (1997).

We simulated a theoretical vaccine with 98% efficacy and coverage of 95% of the susceptible population, an achievable program. The simulated period of vaccination varied from 0 to 70 years and we calculated, through the mathematical model described above, the reduction in the number of liver transplantations carried out each year. Results can be seen in Fig. 11.2, which shows the percentage reduction in the number of liver transplantations as a function of the vaccination time.

Note that the program is entirely inefficient until 20 years of vaccination and its impact rises linearly with time, reaching a maximum of 40% reduction.

It should be remembered that the model assumes approximately 50% of all the liver transplantations carried out in the state of Sao Paulo are due to HCV infection. Therefore, the maximum reduction in the number of

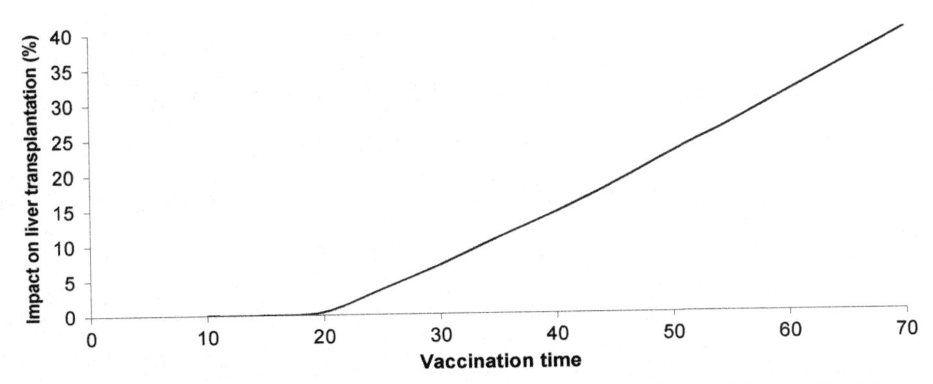

■ **FIGURE 11.2** Percentage reduction in the number of liver transplantations as a function of the vaccination time.

transplantation attained after 70 years is 10% less of the theoretical optimum. This is due to the 2% of primary vaccination failure plus the 5% in the coverage failure, which leaves a small proportion of susceptible individuals who will catch the infection and evolve to liver failure.

Conclusions

HCV-related end-stage cirrhosis is currently the first cause of liver transplantation (Chaib and Massad, 2008a,b,c). The risk for developing cirrhosis 20 years after initial HCV infection among those chronically infected varies between studies, but it is estimated at around 10%–15% for men and 1%–5% for women. Once cirrhosis is established, the rate of developing HCC is at 1%–4% per year (Yu et al., 2009).

Currently, chronic HCV infection–related cirrhosis is the most common indication for liver transplantation in the United States and most parts of the world. While the incidence of new HCV cases has decreased, the prevalence of infection will not peak until the year 2040. In addition, as the duration of infection increases, the proportion of new patients with cirrhosis will double by 2020 in an untreated patient population. If this model is correct, the projected increase in the need for liver transplantation secondary to chronic HCV infection will place an impossible burden on an already limited supply of organs.

In summary, the health care burden caused by hepatitis C is projected to increase significantly in the next 20 years, on the basis of modeling estimates of cirrhosis, hepatic decompensation, and HCC likely to be seen in this population in the future (Lawrence, 2000).

There is currently a great and understandable enthusiasm with the prospects of an anti-HCV vaccine. It should be, therefore, expected that as soon as the vaccine is available, its impact on the number of people developing liver failure and consequently in need of a liver transplantation would be immediately felt. Hence, a dramatic and quick drop on the number of liver transplantation due to HCV infection should be observed as a result of mass vaccination against HCV.

However, due to its rather long development, from infection to liver failure, which may last several decades, people being infected just before the introduction of vaccination would take a long time to develop liver failure. In addition, there are currently an estimated 1 million people harboring HCV in our state. This very large cohort will have a fraction of it that will develop liver failure in many years to come.

We could trace a parallel between this situation and that of Chagas' disease in Brazil. In spite of the fact that its vectorial transmission was officially interrupted in 2006, due to its enormous contingent of infected people, it will take more than 50 years for its complete eradication from the country (Massad, 2008).

Our model confirms our initial hypothesis that in the first 20 years after hepatitis C vaccine development, there will be no change in the liver transplantation demand, and from this moment on, it will take at least 70 years to our population to get rid of the hepatitis C infection and its consequences, subsequently easing down the demand for liver transplantation.

Finally, the advent of a vaccine is welcome and public health authorities should employ a great effort to maximize the number of susceptible individuals immunized. However, decision-makers should be prepared for a very long time living with this scourge.

11.2 DOES ANTI–HEPATITIS B VIRUS VACCINE MAKE ANY DIFFERENCE IN LONG-TERM NUMBER OF LIVER TRANSPLANTATION?

Introduction

It is now more than two millennia since the first recognition of clinical hepatitis (Hilleman, 2011). Hepatitis B virus (HBV) infection is now preventable by a highly effective vaccine that induces long-term memory (Hilleman, 1993). As HBV is a stable virus and has no animal reservoir, it is theoretically eradicable if the vaccine is sufficiently and widely applied for a long time (Hilleman, 2011). As an important cause of liver failure, it should be expected that HBV eradication would have a significant impact on the number of liver transplantation.

Despite the existence of successful vaccine and antiviral therapies, infection with HBV continues to be a major cause of acute and chronic liver disease worldwide. The *sequelae* of HBV infection include acute and chronic infection, cirrhosis of the liver, and HCC.

According to the WHO recommendations, universal vaccination has been currently implemented in 168 countries worldwide with an outstanding record of safety and efficacy. In Brazil, the vaccine was introduced in the national program of immunization only in 2003 (MSB, 2012) and, therefore, the current analysis was based on a nonimmunized population.

A recent research in the Google Scholar database with the terms "impact of vaccine against hepatitis B on liver transplantation" resulted in 23,100 hits. The great majority of studies were related to treatment of reinfection of HBV and/or vaccination after liver transplantation. Hence, to the best of our knowledge, this is the first work to deal with the impact of preventive vaccination of a population at risk to acquire HBV on the number of liver transplantation.

In the absence of any previous study comparing population treated and nontreated with respect to the number of liver failure due to HBV, we have decided to apply a model previously proposed to study the projected impact of vaccination against hepatitis C on liver transplantation (Chaib et al., 2010) to the case of hepatitis B as a cause of liver transplantation.

Methods

The model

As the model presented in the previous chapter, this one was also proposed to study hepatitis C (Massad et al., 2009), without treatment or vaccination, and it assumes a population divided into 14 states. In addition, the model assumes a prevalence of HBV infection of 0.5% and that approximately 20% of all the liver transplantations carried out in the state of São Paulo are due to HBV infection.

Susceptible individuals, denoted $S(t)$, are subdivided into two classes, denoted as $S_1(t)$, representing the general population, and $S_2(t)$, representing a group with higher risk of acquiring the infection by HBV, such as injecting drug users, recipients of blood/blood products transfusion, people occupationally exposed to blood/blood products, unsafe sex, etc. $S_1(t)$ individuals are transferred to the high-risk group $S_2(t)$ with rate ξ. Both susceptible classes acquire the infection with rates β_I ($I = 1, \ldots, 4$) from four different types of individuals, namely, infected and asymptomatic but already infectious individuals, denoted HBV(t), individuals with acute hepatitis, denoted Ac(t), and individuals with intermediate fibrosis stages, called "chronic state," denoted Chr(t). Only a fraction u of β_I is effective for people in the nonrisk group. Susceptible individuals can develop other hepatopathies causing liver failure with rate ω, after which they are transferred to the state called Oth(t). Infected and asymptomatic individuals, HBV(t), can evolve to acute or chronic hepatitis, with rates σ_1 and σ_2, respectively. These individuals evolve to Chr(t) at rate δ_1. Once in the chronic state, individuals can either evolve to HCC, denoted HCC(t), at rate θ, or evolve to liver failure at rate δ_2. We consider only four levels of the MELD, a scale of hepatopathy that incorporates three widely available laboratory variables including the INR, serum creatinine, and serum bilirubin. Those individuals are denoted as MELD$_i$ ($I = 1, \ldots, 4$). Evolution between the MELD levels occurs at rates ε_I ($I = 1, \ldots, 3$). Infected individuals without chronic hepatitis can recover to compartment $R(t)$ at rates γ_I ($I = 1, \ldots, 4$). Individuals from compartments MELD$_i$ ($I = 1, \ldots, 4$) and HCC(t) get into the waiting list for liver transplantation, WL(t), at rates φ_I ($I = 1, \ldots, 6$). Individuals in the waiting list are eventually transplanted at rate τ_1. We also consider the possibility of lost of graft at rate φ_6 and a secondary waiting list WLTx(t). From the latter, individuals are eventually transplanted at rate τ_2. Every individual in this population is subjected to a mortality rate μ and an additional mortality of those with liver disease occurs at rates α_I ($I = 1, \ldots, 10$) depending on the state. Finally, the susceptible compartment grows at rate Λ, which is comprised by the sum of all mortalities to keep the total population constant.

The model's structure is summarized in Fig. 11.3.

The model's dynamics is described by the following set of equations:

$$
\begin{aligned}
\frac{dS_1(t)}{dt} &= \Lambda - u\beta_1 HBV(t)\frac{S_1(t)}{N} - u\beta_2 Ac(t)\frac{S_1(t)}{N} - u\beta_3 T_A(t)\frac{S_1(t)}{N} \\
&\quad - u\beta_4 Chr(t)\frac{S_1(t)}{N} - (\mu + \omega + \xi)S_1(t) \\
\frac{dS_2(t)}{N} &= \xi S_1(t) - \beta_1 HBV(t)\frac{S_2(t)}{N} - \beta_2 Ac(t)\frac{S_2(t)}{N} - \beta_3 T_A(t)\frac{S_2(t)}{N} \\
&\quad -\beta_4 Chr(t)\frac{S_2(t)}{N} - (\mu + \omega)S_2(t) \\
\frac{dHBV(t)}{dt} &= \beta_1 HBV(t)\frac{(S_1(t) + S_2(t))}{N} + \beta_2 Ac(t)\frac{(S_1(t) + S_2(t))}{N} \\
&\quad +\beta_3 T_A(t)\frac{(S_1(t) + S_2(t))}{N} + \beta_4 Chr(t)\frac{(S_1(t) + S_2(t))}{N} \\
&\quad -(\gamma_1 + \sigma_1 + \sigma_2 + \mu)HBV(t)
\end{aligned}
$$

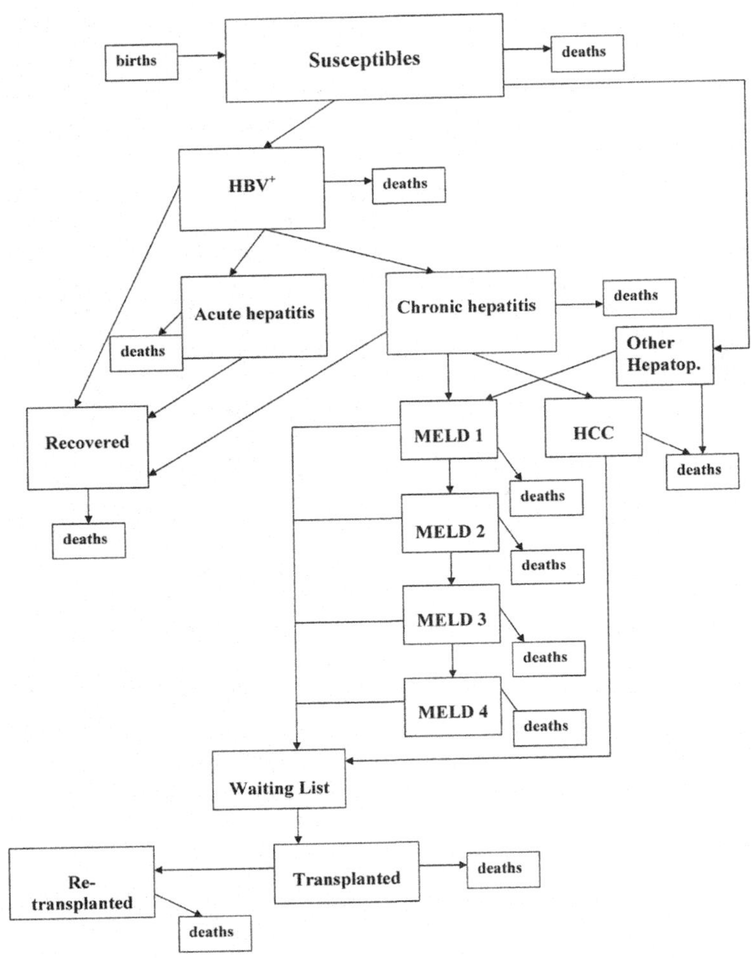

■ **FIGURE 11.3** Model's structure according to the natural history of hepatitis B and other hepatic diseases and organ transplantation. *HBV*, hepatitis B virus; *HCC*, hepatocellular carcinoma; *MELD*, Model for End-Stage Liver Diseases. *Adapted from Massad, E., Coutinho, F.A.B., Chaib, E., Burattini, M.N., 2009. Cost-effectiveness analysis of a hypothetical hepatitis C vaccine as compared to antiviral therapy. Epidemiol. Infect. 137 (2), 241–249.*

$$\frac{dAc(t)}{dt} = \sigma_1 HBV(t) - (\delta_1 + \gamma_2 + \alpha_1 + \mu)Ac(t)$$

$$\frac{dR(t)}{dt} = \gamma_1 HBV(t) + \gamma_2 Ac(t) + \gamma_4 Chr(t) - \mu R(t)$$

$$\frac{dChr(t)}{dt} = \delta_1 Ac(t) + \sigma_2 HBV(t) - (\delta_2 + \theta + \alpha_2 + \gamma_3)Chr(t)$$

$$
\begin{aligned}
\frac{dMELD_1(t)}{dt} &= \delta_2(Chr(t) + Oth(t)) - (\varphi_1 + \varepsilon_1 + \alpha_3 + \mu)NT_CM_1(t) \\
\frac{dMELD_2(t)}{dt} &= \varepsilon_1 NT_CM_1(t) - (\varphi_2 + \varepsilon_2 + \alpha_4 + \mu)NT_CM_2(t) \\
\frac{dMELD_3(t)}{dt} &= \varepsilon_2 NT_CM_2(t) - (\varphi_3 + \varepsilon_3 + \alpha_5 + \mu)NT_CM_3(t) \\
\frac{dMELD_4(t)}{dt} &= \varepsilon_3 NT_CM_3(t) - (\varphi_4 + \alpha_6 + \mu)NT_CM_4(t) \\
\frac{dHCC(t)}{dt} &= \theta Chr(t) - (\varphi_5 + \alpha_7 + \mu)HCC(t) \\
\frac{dWL(t)}{dt} &= \sum_{i,j=1}^{4} \varphi_i NT_CM_j(t) + \varphi_5 Chr(t) - (\tau_1 + \alpha_8 + \mu)WL(t) \\
\frac{dTx_1(t)}{dt} &= \tau_1 WL(t) - (\varphi_6 + \alpha_9 + \mu)Tx_1(t) \\
\frac{dWLTx(t)}{dt} &= \varphi_6 Tx_1(t) - (\tau_2 + \alpha_{10} + \mu)WLTx(t) \\
\frac{dTx_2(t)}{dt} &= \tau_2 WLTx(t) - (\alpha_{11} + \mu)Tx_2(t) \\
\frac{dOth(t)}{dt} &= \omega(S_1(t) + S_2(t)) - (\delta_2 + \alpha_{12} + \mu)Oth(t)
\end{aligned}
\tag{11.2}
$$

The biological meaning and the values of the parameters can be seen in Table 11.2.

Table 11.2 Parameters used in the simulations.

Parameters	Biological meaning	Numerical values
Λ	Births	Variable
u	Attenuation of β_i affecting people in the nonrisk group	0.05
ξ	Rate of transfer to high-risk group	1.0×10^{-5} days^{-1}
μ	Natural mortality rate	3.9×10^{-5} days^{-1}
β_i	Potentially infective contact rates	5.0×10^{-4} days^{-1}
γ_i	Recovery rates from infection	From 5.0×10^{-5} to 1.0×10^{-4} days^{-1}
σ_i	Rates of evolution to hepatitis	From 2.4×10^{-3} to 3.0×10^{-3} days^{-1}
δ_1	Rates of evolution from acute to chronic hepatitis	8.4×10^{-4} days^{-1}
δ_2	Rate of evolution to hepatic failure (MELD)	6.0×10^{-5} days^{-1}
θ	Rate of evolution of hepatocellular carcinoma	1.36×10^{-6} days^{-1}
φ_i	Rates of getting into the waiting lists	From 1.1×10^{-7} to 2.0×10^{-6} days^{-1}
τ_i	Transplantation rates	Variable
ω	Rate of evolution to other hepatopathies	1.56×10^{-6} days^{-1}
α_i	Disease-induced mortality rates	From 4.2×10^{-5} to 5.9×10^{-5} days^{-1}

Results

Model's simulation

We simulated a theoretical vaccine with 98% efficacy and coverage of 95% of the susceptible population, an achievable program. The simulated period of vaccination varied from 0 to 70 years and we calculated, through the mathematical model described above, the reduction in the number of liver transplantations carried out each year. Results can be seen in Fig. 8.2, which shows the percentage reduction in the number of liver transplantations as a function of the vaccination coverage. In the figure, we show different levels of HBV infection prevalence, varying from 0.1% to 2.5%. We see that the higher the equilibrium prevalence of HBV, the higher the impact of vaccination on the number of liver transplantation.

The impact of a hypothetical hepatitis B vaccine on the liver transplantation program. Continuous line represents the epidemiological situation of HBV in São Paulo, Brazil. Dotted lines are simulations of different scenarios.

Note that the program is entirely inefficient, and its impact reaches a maximum of about 14% reduction.

It should be remembered that the model assumes that, approximately, 20% of all the liver transplantations carried out in the state of São Paulo are due to HBV infection.

Conclusions

HBV infection is still a major global health problem, despite the first vaccine being made available in 1981 (Luo et al., 2011; Zhang et al., 2011).

About 2 billion people worldwide have been infected with the virus, an estimated 360 million live with chronic infection, and at least 600,000 people die annually from acute or chronic consequences of hepatitis B.

Currently, HBV causes nearly 1 million deaths per annum globally. HBV vaccination will have a major impact on prevalence of infection. HBV is highly prevalent in Asia, Africa, and parts of Southern and Eastern Europe.

Vaccination plays a central role in HBV prevention strategies worldwide, and a decline in the incidence and prevalence of HBV infection following the introduction of universal HBV vaccination programs has been observed in many countries, including the United States and parts of Southeast Asia and Europe. Therefore, vaccination is the most effective way to achieve global HBV control and prevent cirrhosis, liver failure, and liver cancer.

For instance, in the United Kingdom, more than 300,000 people may have chronic HBV infection (Hepatitis B Foundation UK, 2007); however, HBV-related diseases' indication for transplantation is in only 5% of transplant cases; it probably means that widespread vaccination makes HBV not a leading indication for liver transplantation. Otherwise, in our study, 20% of liver transplantations are due to HBV chronic liver disease because vaccination is not a compulsory program as part of routine universal childhood vaccination schedule.

Liver transplantation, in recent years, has become the treatment of choice for patients with end-stage liver disease. Chronic liver disease due to hepatitis B represents 5%—10% of all liver transplantations performed in Europe. The prognosis after the surgery is related to the efficacy of prophylaxis of HBV graft reinfection. The risk of HBV reinfection is directly related to the HBV viral load at transplantation.

The use of mathematical models in transplantation issues, notwithstanding its apparent oddness, has already provided insights into the important questions related to the understanding of complex questions related to causes of organ failures and transplantation policies (de Carvalho et al., 1996; Chaib and Massad, 2005, 2008a,b,c; Chaib and Massad, 2007a; Chaib and Massad, 2007b).

The model here applied is an adaptation of previous models used for the quantification of the dynamics of HCV (de Carvalho et al., 1996) and of liver transplantation in HCV-infected patients, as well as the impact of a potential vaccine on the number of transplantations in a hypothetical population (Chaib and Massad, 2007).

Our model assumes that, approximately, 20% of all the liver transplantations carried out in the state of São Paulo are due to HBV infection. This high proportion is probably due to the low vaccination coverage against HBV in São Paulo. Other epidemiological scenarios would result in different vaccination impact. So, for instance, if the prevalence of HBV was higher than the 5% observed for the state of São Paulo, the contribution of HBV infection to the total number of liver transplantation would be higher than the 20%. In this case, the impact of vaccination would be higher than the one we estimated (Fig. 11.5). This is because the relative contribution of other causes of liver failure would be relatively less important. Our results suggest that a vaccination program that covers 80% of the target population would reach a maximum of about 14% reduction in the liver transplantation program.

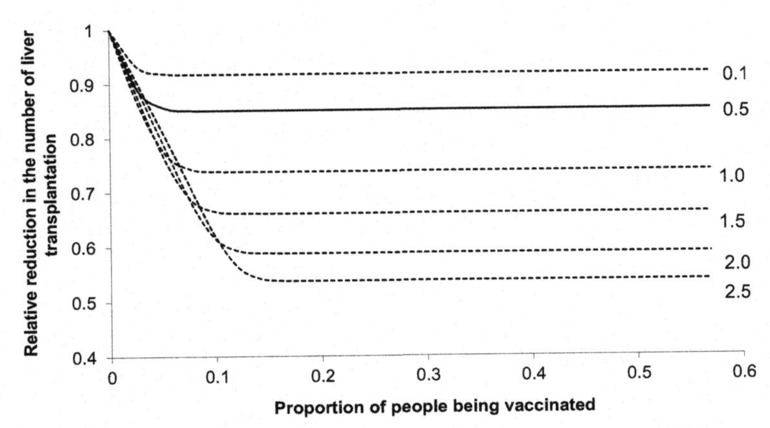

■ **FIGURE 11.4** Percentage reduction in the number of liver transplantations as a function of the vaccination coverage.

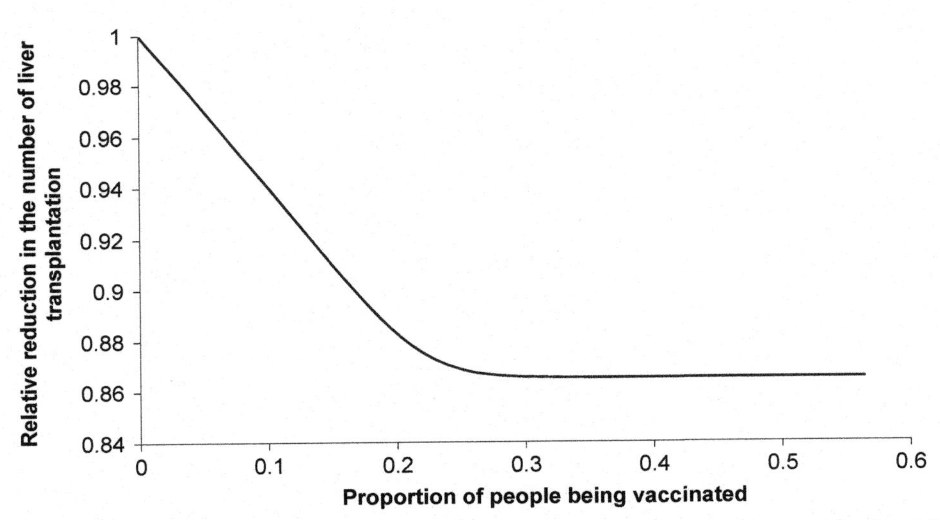

■ **FIGURE 11.5** A theoretical scenario in which the prevalence of hepatitis B virus is 20%. In this case, the impact of vaccination is substantially higher than in the ones shown in Fig. 11.2.

This relatively low impact is probably because of the fact that the transplantation program in the state of São Paulo is overwhelmed by the high number of patients with liver failure due to other causes, mainly chronic hepatitis C. Hence, the number of liver failures avoided by HBV vaccination would be quickly offset by other nosological entities. In Fig. 11.5, we show a theoretical scenario in which the prevalence of HBV is 20%. In this case, the impact of vaccination is substantially higher than in the ones shown in Fig. 11.4.

Also noteworthy is the fact that the vaccination impact plateau at vaccination coverage above 75%.

The impact of a hypothetical hepatitis B vaccine on the liver transplantation program for the theoretical scenario of 20% of HBV prevalence.

Finally, it should be stressed that, in spite of the poor impact of vaccination against hepatitis B on the expected number of liver transplantation, HBV vaccination programs continue to be fully justified. We are far from suggesting that the HBV vaccination program is not worthwhile doing.

In conclusion, our analysis suggests that increasing the vaccination coverage against HBV in the state of São Paulo would have a relatively low impact on the number of liver transplantation. In addition, this impact would take several decades to materialize because of the long incubation period of liver failure due to HBV.

Impact of HCV antiviral therapy on the liver transplantation waiting list assessed by mathematical models

12.1 INTRODUCTION

Prevalence of hepatitis C virus (HCV) infection is found worldwide, however, country prevalence ranging from less than 1% to greater than 10%. The highest prevalence has been reported in Africa and Middle East, with a lower prevalence in the Americas, Australia, and Northern and Western Europe (Hajarizadeh et al., 2013).

As mentioned in the previous chapters, since the first liver transplantation (LT) in the State of São Paulo (Machado, 1972), the recipient waiting list has increased; now approximately 150 new cases per month are referred to the single list at the central organ procurement organization (Chaib and Massad, 2005).

HCV infection is considered a major public health problem (Petruzziello et al., 2016) with a global prevalence rate of 2.8%, equating to over 185 million infections and more than 350,000 deaths annually.

An estimated 3 million to 4 million new cases of HCV infection emerge every year worldwide (Chaib and Massad, 2008a,b,c). Furthermore, the HCV-related mortality is increasing and HCV infection is projected to be the most important leading cause of viral hepatitis-related mortality in the near future (Morgan et al., 2013; Petruzziello et al., 2016).

End-stage liver disease due to HCV is currently the leading indication for LT in both the United State of America and Brazil, mainly in the State of Sao Paulo accounting for over 30% and 40% of all transplants annually, respectively (Crespo et al., 2012; Morgan et al., 2013). However, treatment for chronic HCV infection, with elimination of HCV infection, has

Mathematical Approaches to Liver Transplantation. https://doi.org/10.1016/B978-0-12-817436-4.00012-6

revolutionized in the past 5 years with the approval of second-generation direct-acting antiviral agents.

The number of patients on the liver transplantation waiting list (LTWL) in the State of Sao Paulo jumped 2.71-fold in the past 10 years, almost 50% of them due to HCV; consequently the number of deaths on LTWL moved to a higher level increasing 2.09-fold (Massad et al., 2009).

Our aim is to analyze, through a mathematical model, the potential impact of HCV antiviral therapy on the LTWL in the State of Sao Paulo, Brazil.

12.2 MATERIALS AND METHODS

This is a theoretical work and we used mathematical models designed to mimic the LTWL's dynamics and which represent improvements on works previously published. In previous papers, Chaib et al. (2005; 2007; 2008a and 2008b) projected the size of the waiting list, L, by taking into account the incidence of new patients per year, I, the number of transplantations carried out in that year, Tr, and the number of patients who died in the waiting list, D. As shown in Chapter 5, the dynamics of the waiting list is given by the recurrent equation $L_{t+1} = L_t + I_t - D_t - Tr_t$, that is, the list size at time $t+1$ is equal to the size of the list at the time t, plus the new patients getting into the list at time t, minus those patients who died in the waiting list at time t, and minus those patients who received a graft at time t. The variables I and D from 2006 onward were projected by fitting an equation by maximum likelihood, in the same way that we did for Tr.

In this chapter, we improved the list dynamics by considering a continuous time model as follows:

$$\frac{dL(t)}{dt} = (\beta - \alpha - Tr)L(t) \qquad (12.1)$$

where β is the incidence rate of patients with the Model for End-Stage Liver Disease criteria to get into the LTWL, α is the death rate, and T_r is the transplantation rate of patients in the LTWL, respectively. We used the Latin hypercube sampling method to find the values of the parameters that would explain the observed data.

Eq. (12.1) has the following solution:

$$L(t) = \exp[-(\beta - \alpha - Tr)t] \qquad (12.2)$$

Data used in the work have been collected in the Service of Transplantation of the State Secretary of Health of Sao Paulo.

12.3 RESULTS

Table 12.1 shows the value of the variables that entered the model. From the time variation in each of the variables, we estimated the rates of Eq. (12.1).

Fig. 12.1 shows a comparison between the actual number of patients in the LTWL from 2006 until 2017, an exponential fitting, and the integral of Eq. (12.1).

As can be observed in the figure, the set of parameters used retrieves the actual data with the same accuracy as the exponential fitting. This should be expected because the solution of equation $L_{t+1} = L_t + I_t - D_t - Tr_t$ is also an exponential function (Eq. 12.1). However, the remarkable tally of the models output with the exponential fitting was obtained by optimizing the value of the parameters through the Latin hypercube sampling technique used.

We projected the result of Eq. (12.2) for the next 30 years under the assumption that all the conditions would remain the same. Next, we introduced the anti-HCV treatment, which was assumed to halve the incidence of patients in the LTWL and that the recovery of patients in the list would triple. The LT

Table 12.1 Value of the variables that entered into the model.

Year	Previous No. of patients in the liver transplantation waiting list	Incidence of new cases per year	Number of transplants	Recovered	Deaths
1	4183	1565	510	24	840
2	4013	1022	440	44	708
3	4165	1212	545	54	461
4	4266	1287	693	45	448
5	4544	1415	744	37	356
6	5023	1576	688	15	394
7	5537	1490	599	18	359
8	5998	1520	653	12	394
9	6262	1378	665	16	433
10	6643	1451	655	19	396
11	6916	1316	623	21	399
12	7181	1385	620	28	472

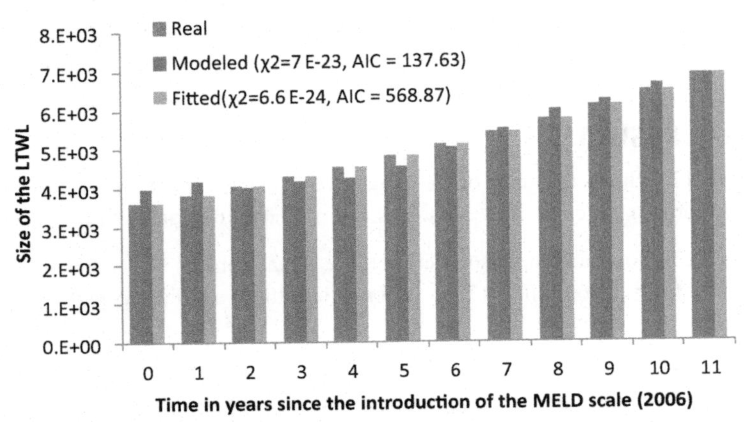

■ **FIGURE 12.1** Number of patients in the liver transplantation waiting list (LTWL) since 2006. Real data (blue *[dark gray in print version]*) are compared with the simple fitting procedure (green *[light gray in print version]*) and the model (red *[black in print version]*).

rate was assumed to not be affected by the anti-HCV treatment. Fig. 12.2 shows the results of this simulation.

It can be seen from Fig. 12.2 that the anti-HCV treatment would have a remarkable impact on the size of the LTWL, dropping from around 24,000 patients to around 12,000 patients in 30 years. This reduction would obviously be more accentuated if the transplantation capacity of the system would increase, such that by doubling the transplantation rate from year 12, the LTWL would be reduced to zero after less than 10 years.

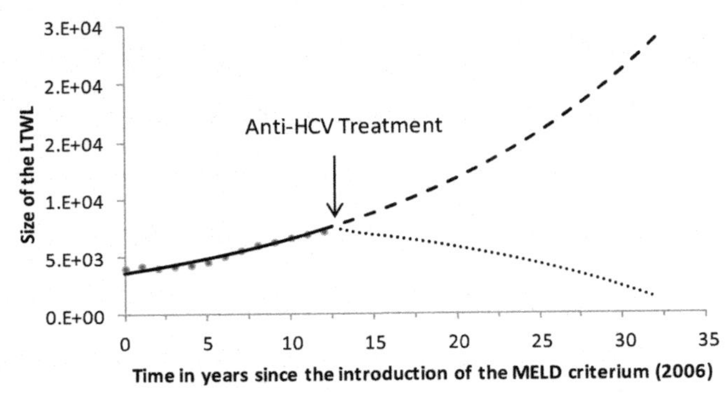

■ **FIGURE 12.2** Impact of anti—hepatitis C virus (HCV) treatment on the size of the liver transplantation waiting list (LTWL). Red dots *(black in print version)* represent real data, gross-dotted line represents the projection of the size of the LTWL in the absence of treatment, and finely *dotted line* represents the theoretical reduction in the size of the LTWL as a result of treatment introduced at time 12.5 years.

12.4 DISCUSSION

In this work we propose an improvement of mathematical models aimed to mimicking the LTWL's dynamics previously published to assess the theoretical impact of anti-HCV treatment on the size of the list. It represents an important contribution to the understanding of the possible impact of anti-HCV treatment on the number and rate of new liver failure cases in an affected community because HCV-related end-stage cirrhosis is currently the first cause of LT (Chaib and Massad, 2008b). The risk for developing cirrhosis 20 years after initial HCV infection among those chronically infected varies between studies, but it is estimated at around 10%–15% for men and 1%–5% for women. Once cirrhosis is established, the rate of developing hepatocellular carcinoma (HCC) is at 1%–4% per year.

The health care burden caused by hepatitis C is projected to increase significantly in the next 20 years, on the basis of modeling estimates of cirrhosis, hepatic decompensation, and HCC likely to be seen in this population in the future (Lawrence, 2000).

Currently, chronic HCV infection–related cirrhosis is the most common indication for LT in the United States and most parts of the world. While the incidence of new HCV cases has decreased, the prevalence of infection will not peak until the year 2040. In addition, as the duration of infection increases, the proportion of new patients with cirrhosis will double by 2020 in an untreated patient population. In previous papers, our group (Chaib et al. 2010, 2012, 2013; Amaku et al., 2013, 2017; Chaib et al., 2013a,b) proposed a series of mathematical models dealing with distinct aspects of LT. Some of the models were simple like the present one and some more complex. In spite of the mathematical simplicity of the current model, it may serve as a benchmark to test a crucial hypothesis related to the size of the waiting list in countries with a rather limited supply of grafts, namely, how a new tool of public health control (in this case a new therapy) will impact the rate at which liver failure patients could be transplanted as soon as possible. We think that the present simple model can represent an important step forward in understanding and controlling this rather important public health issue.

Finally, it is important to highlight the model's limitations such as the simplicity of the assumptions behind the equations. However, the role of mathematical models as applied to real-world problems consists in helping the understanding of the phenomenon and providing tools of predictive capacity that may be used to guide decision-making, in particular in critical health issues such as LT. We hope that this model may be of clinical use related to the optimal distribution of anti-HCV treatment as a control tool to end-stage liver failure.

12.5 CONCLUSION

Our mathematical model demonstrates that anti-HCV therapy would have a remarkable impact on the size of the LTWL, in the State of Sao Paulo, dropping from 24000 to approximately 1200 patients in the next 30 years.

Future perspectives

13.1 THE FUTURE OF THE MATHEMATICS OF LIVER TRANSPLANTATION

Modeling an optimum MELD score that would minimize mortality

In this final section, we propose a model for minimizing the total mortality in the LTWL that could determine what would be the optimum MELD score to matriculate individuals in the list. The model follows the same step as the one presented in Chapter 6 for optimizing the size of liver tumors of candidate to liver transplantation. We used the data from the Secretary of Health of the State of São Paulo, Brazil, which lists 22,522 patients for the period between 2006 (when the MELD score was introduced in Brazil) and June 2019. The model is described in this final chapter on future perspectives by reasons that will be explained along the text.

We begin by assuming that patients with liver failure present themselves along a short time interval ΔT with MELD scores of variables magnitudes. As in the case of the liver tumor model, we call this interval "at presentation." During this time interval, we assumed that N liver failure patients are included in the transplantation waiting list, and that F livers are available to these patients. Note that we use the same notation as in the model for liver tumors presented in Chapter 6.

We consider the mortality of nontransplanted and transplanted patients as a function of the MELD score at presentation. Fig. 13.1 shows the probability density function of the MELD score of those 22,552 patients at presentation.

As it can be observed, the MELD score at presentation has an exponential distribution with parameter $\lambda = 0.05$, which means that on average, patients are matriculated in the LTWL with MEDL score equals to 20.

Mathematical Approaches to Liver Transplantation. https://doi.org/10.1016/B978-0-12-817436-4.00013-8

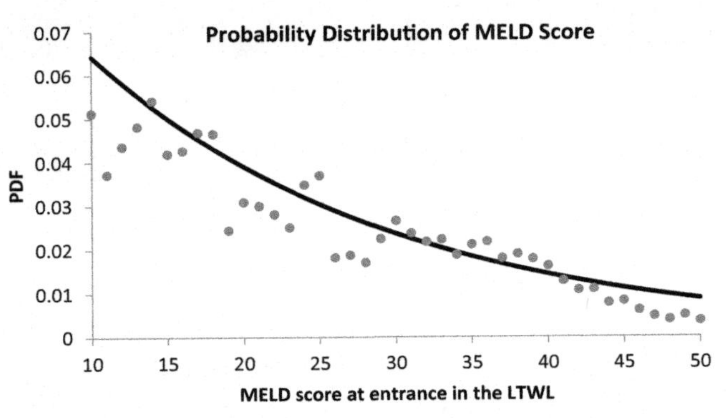

■ **FIGURE 13.1** Probability distribution of MELD scores at presentation.

Mortality in the LTWL for nontransplanted and transplanted patients

Among the 22,552 patients listed in the LTWL from 2006 to 2019, 6121 were transplanted, and 16,431 were not transplanted. Of the transplanted individuals, 2401 died in the period, whereas of the 16,431 nontransplanted, 4779 died in the list. This represents a total mortality of 39.2% for transplanted and 29% for nontransplanted patients. We applied the Chi-square test to compare the significance of the above difference, which resulted in $\chi^2 = 195.667$ with $p < 0.00001$. This higher mortality rate among transplanted than nontransplanted patients needs further investigation, and this is the reason this analysis is presented in the future perspective chapter.

The survival of both groups of patients along the 16 years of the analysis is shown in Figs. 13.2 and 13.3 as a function of the MELD score at presentation for the nontransplanted and transplanted patients, respectively.

As can be noted in the figures above, there is no difference between the two groups (Mann–Whitney U test = 11,777, p = 0.56).

Next we calculated the probability of death for both groups along the 16 years of analysis as a function of MELD score at presentation. Figs. 13.4 and 13.5 show the results for the nontransplanted and transplanted patients, respectively.

As can be noted, the form of the curves is entirely different from each other. The probability of death of nontransplanted patients growths with logarithm way, whereas the probability of death of transplanted patients growths in an exponential fashion.

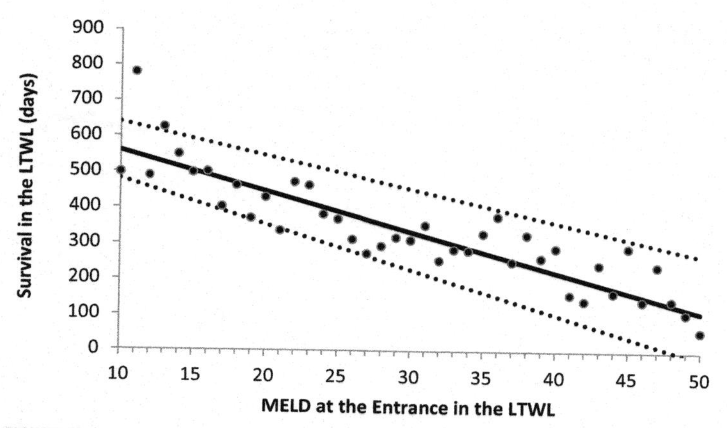

■ **FIGURE 13.2** Survival in the LTWL of nontransplanted patients as a function of MELD at presentation. *Dots* represent real data, *solid line* the average fitting, and *dotted lines* the 95% C.I.

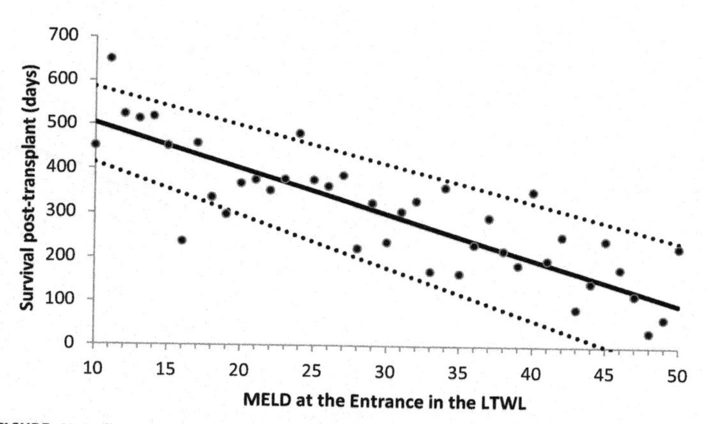

■ **FIGURE 13.3** Survival of transplanted patients as a function of MELD at presentation. *Dots* represent real data, *solid line* the average fitting, and *dotted lines* the 95% C.I.

By the same token as for the case of liver tumors, the model is based on four assumptions, namely,

1. the mortality rate of nontransplanted, α_{nt}, and transplanted, α_t, liver failure patients is calculated from the actual mortality probabilities according to the following equations:

$$\alpha_{nt}(s) = \alpha_0/s \tag{13.1}$$

and

$$\alpha_t(s) = \beta e^{\delta s} \tag{13.2}$$

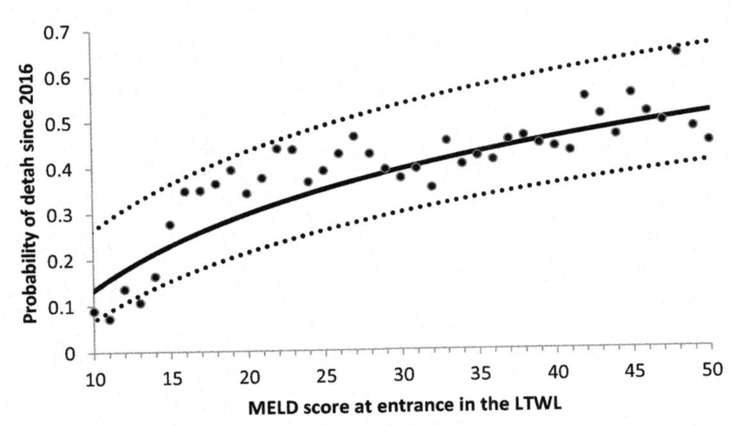

■ **FIGURE 13.4** Death probability in the LTWL of nontransplanted patients as a function of MELD at presentation. *Dots* represent real data, *solid line* the average fitting, and *dotted lines* the 95% CI.

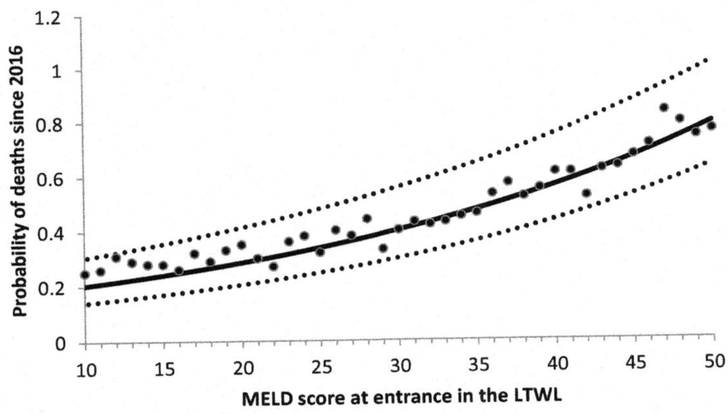

■ **FIGURE 13.5** Death probability of transplanted patients as a function of MELD at presentation. *Dots* represent real data, *solid line* the average fitting, and *dotted lines* the 95% CI.

where s is the MELD score at presentation and α, δ, and β are the parameters obtained from the fitting of Figs. 13.4 and 13.5. In Eqs. (13.1) and (13.2), take into account the fact that MELD scores increase with time and so does the mortality rates. Eqs. (13.1) and (13.2) are illustrated in Fig. 13.6, in which the mortality rates for both the transplanted and nontransplanted patients as a function of the MELD score, s, at presentation are shown.

The probability of surviving after T years for nontransplanted and transplanted patients, $\pi_{nt}(s)$ and $\pi_t(s)$, respectively, as a function of their

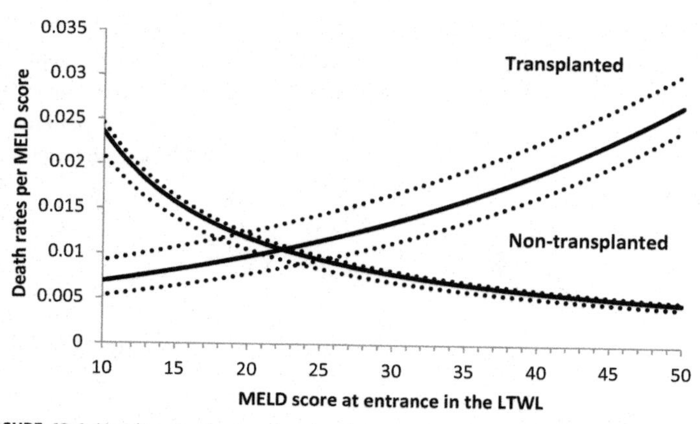

FIGURE 13.6 Mortality rates for transplanted and nontransplanted patients as a function of MELD score at presentation. Continuous lines represent average and *dotted lines* the respective 95% CI.

MELD score, s, at the time individuals are included in the transplantation program, is given by

$$\pi_{nt}(s) = \exp(-\alpha_{nt}T) \tag{13.3}$$

and

$$\pi_t(s) = \exp(-\alpha_t T) \tag{13.4}$$

Eqs. (13.3) and (13.4) result in survival probabilities after T years that are in agreement with the real data, as shown in Fig. 13.1. They were used to calculate the form and parameters of Eqs. (13.1) and (13.2).

2. the mortality of both transplanted and nontransplanted patients is a monotonically increasing function of MELD score at presentation, as shown in Figs. 13.4 and 13.5 (MELD score is, therefore, taken as an indication of gravity);
3. the number of available livers to be grafted, F, is limited and always less than the total number of liver failure patients, N, who have transplantation indication;
4. finally, the MELD score, s, at the time individuals are included in the transplantation program, is distributed in the liver failure population according to an exponential distribution, as shown in Fig. 13.1, according to the following equation:

$$f(s, \lambda) = \lambda e^{\lambda s} \tag{13.5}$$

where λ is the *rate parameter* of the distribution. This implies that in a liver failure population, many individuals have MELD score of small magnitude and a very low number of who present MELD score of larger magnitude. Again, this distribution of MELD score is at the moment the patients get into the transplantation program. The *cumulative distribution function* is given by

$$F(s, \lambda) = \int_0^s \lambda e^{\lambda t} dt = 1 - e^{\lambda s} \tag{13.6}$$

Eq. (13.6) means the probability that a given liver failure patient has MELD score equal or less than s.

The exponential distribution has mean (*expected value*) equal to

$$E[s] = \frac{1}{\lambda} \tag{13.7}$$

and variance

$$Var[s] = \frac{1}{\lambda^2} \tag{13.8}$$

With the above assumptions, we define $p(s)ds$ as the proportion of individuals with MELD score between s and $s+ds$; $x(s)ds$ the proportion of transplanted patients with MELD score between s and $s+ds$; and $y(s)ds$ the proportion of nontransplanted patients with MELD score between s and $s+ds$. These proportions are related such that

$$x(s) = \frac{F}{N} p(s) \tag{13.9}$$

and

$$y(s) = \left(1 - \frac{F}{N}\right) p(s) \tag{13.10}$$

Eqs. (13.9) and (13.10) can be interpreted as follows: a proportion $p(s)$ of the patients has MELD score s, of which a fraction $\frac{F}{N}$ is transplanted and its complement $\left(1 - \frac{F}{N}\right)$ is not transplanted, such that $x(s) + y(s) = p(s)$. Note that, this was a particular transplantation policy. For example, we could replace Eqs. (13.9) and (13.10) by $x(s) = g(s)\frac{F}{N}p(s)$ and $y(s) = \left(1 - g(s)\frac{F}{N}\right)p(s)$, where $g(s)$ is some bias toward any eventual MELD score preference. In this work, $g(s) = 1$, meaning that all patients have the same chance of being transplanted (no bias). According to adopted criteria,

$$g(s) = \begin{cases} 1 & \text{if } s > s_M = 25 \\ 0 & \text{if } s \leq s_M = 25 \end{cases}$$

We then calculated the following:

1. If we choose to transplant every patient with any MELD score equal or greater than a critical score, S_F, then to guarantee that all patients with such score greater than S_F are transplanted (that is, all grafts are used), S_F has to be defined as

$$N \int_0^{s_F} p(s)ds = F \tag{13.11}$$

or

$$s_F = -\frac{\log\left(1 - \frac{F}{N}\right)}{\lambda} \tag{13.12}$$

In other words, this means to choose a policy such that $x(s \leq s_F) = p(s)$ and $y(s \leq s_F) = 0$.

Eqs. (13.11) and (13.12) can be interpreted as follows: $\int_0^{s_F} p(s)ds$ is the fraction of the population that has MELD score equal or greater than S_F. Multiplied by the total population, N gives the total number of individuals that are transplanted, that is, received all the liver grafts F. In other words, all available livers are used. The size limit that guarantees that this happens, S_F, is therefore calculated as a function of F as in Eq. (13.12).

2. Hence, if not all patients with MELD score s are transplanted, for example, if we choose to transplant $x(s) = \frac{F}{N}p(s)$ and not transplant $y(s) = \left(1 - \frac{F}{N}\right)p(s)$, then we can choose to transplant all the patients with MELD score of $s_0 > s_F$.
3. Using the adopted criteria (see above), the proportion of nontransplanted patients with MELD score s above S_M with respect to the total number of patients at presentation is

$$p_{nt}(s > s_M) = \begin{cases} \dfrac{N(1 - e^{-\lambda s_M}) - F}{N} & \text{if } F > N(1 - e^{-\lambda s_M}) \\ 0 & \text{otherwise} \end{cases} \tag{13.13}$$

Eq. (13.13) means that multiplying the proportion of patients with MELD score equal or greater than S_M, $(1-e^{-\lambda S_M})$ by the total population of patients, N, gives the number of patients with score greater than S_M. This number minus the number of available livers divided by the total population size gives the proportion of nontransplanted patients.

4. The proportion of transplanted patients with respect to the total number of patients at presentation, with tumor size s above S_M, is given as

$$p_t(s > s_M) = \begin{cases} \dfrac{F}{N} & \text{if } F > N(1 - e^{-\lambda s_M}) \\ \dfrac{F}{N}(1 - e^{-\lambda s_M}) & \text{otherwise} \end{cases} \tag{13.14}$$

Eq. (13.14) reflects the fact that a fraction $\frac{F}{N}$ of those individuals with MELD score equal or greater than s_M is transplanted when the number of available livers F is less than the number of individuals with MELD score greater than s_M at presentation.

5. If the adopted criteria are obeyed, then the proportion of nontransplanted patients with MELD score greater than s_M is

$$p_{nt}(s > s_M) = e^{-\lambda s_M} \tag{13.15}$$

which is the minimum (if F is not enough to transplant up to s_M) proportion of individuals with MELD score greater than s_M. According to the adopted criteria, none of those patients are transplanted, independently of F.

6. If the adopted criteria are not obeyed, then there is a proportion of transplanted patients with score less than s_M that could be transplanted. This proportion is limited by the number of available livers, and it is

$$p_t(s < s_M) = \begin{cases} \dfrac{Fe^{-\lambda s_M}}{N} & \text{if } F < N(1 - e^{-\lambda s_M}) \\ 0 & \text{otherwise} \end{cases} \tag{13.16}$$

In this situation, the proportion of nontransplanted is given by

$$p_t(s < s_M) = \begin{cases} e^{-\lambda s_M} & \text{if } F < N(1 - e^{-\lambda s_M}) \\ 1 - \dfrac{F}{N} & \text{otherwise} \end{cases} \tag{13.17}$$

Note that adding the proportion of nontransplanted individuals with scores greater and less than s_M gives $1 - \frac{F}{N}$. By the same token, adding the proportion of transplanted individuals with scores greater and less than s_M gives $\frac{F}{N}$.

7. Now, we abandon the adopted criteria and transplant a proportion $x(s) = \frac{F}{N}p(s)$ of individuals with MELD score less than s_0 (variable) and compare the impact on the total mortality of patients with the mortality resulting from adopting the current criteria.

First, we calculate the survival of transplanted patients (*TS*) with score up to s_0 at a moment in time *T* after the patients' presentation. The proportion of the individuals with tumor size up to s_0 at presentation is

$$\int_0^{s_0} \lambda e^{-\lambda s} ds \qquad (13.18)$$

The proportion of patients at presentation who were transplanted and survived up to *T* after the transplantation is

$$\int_0^{s_0} x(s) e^{-\alpha_t(s)T} ds = \frac{F}{N} \int_0^{s_0} \lambda e^{-\lambda s} e^{-\alpha_t(s)T} ds \qquad (13.19)$$

Hence, the total number of transplanted patients (*TS*) with score size up to S_0 at presentation and who survived up to time *T* is given by Eq. (13.19) multiplied by *N*:

$$TS = N\frac{F}{N} \int_0^{s_0} \lambda e^{-\lambda s} e^{-\alpha_t(s)T} ds = F \int_0^{s_0} \lambda e^{-\lambda s} e^{-\alpha_t(s)T} ds \qquad (13.20)$$

8. The number of patients with score loess than S_0 at presentation who were not transplanted is

$$N \int_0^{s_0} y(s) ds = N \int_0^{s_0} \left(1 - \frac{F}{N}\right) \lambda e^{-\lambda s} ds \qquad (13.21)$$

and those who survived after time *T* are

$$N \int_0^{s_0} \left(1 - \frac{F}{N}\right) \lambda e^{-\lambda s} e^{-\alpha_{nt}(s)T} ds \qquad (13.22)$$

Now, the number of patients with MELD score greater than score s_0 at presentation that were not transplanted is

$$N \int_{s_0}^{\infty} p(s) ds = N \int_{s_0}^{\infty} \lambda e^{-\lambda s} ds \qquad (13.23)$$

and, among those, the survivors after time T are

$$N\int_{s_0}^{\infty} \lambda e^{-\lambda s} e^{-\alpha_{nt}(s)T} ds \quad (13.24)$$

Hence, the total number of survivors after time T who were not transplanted is

$$NTS = N\int_{0}^{s_0} \left(1-\frac{F}{N}\right) \lambda e^{-\lambda s} e^{-\alpha_{nt}(s)T} ds + N\int_{s_0}^{\infty} \lambda e^{-\lambda s} e^{-\alpha_{nt}(s)T} ds \quad (13.25)$$

9. Therefore, the total survival is obtained by adding Eqs. (13.20) and (13.25):

$$\text{Survivors} = F\int_{0}^{s_0} \lambda e^{-\lambda s} e^{-\alpha_{t}(s)T} ds + N\int_{0}^{s_0} \left(1-\frac{F}{N}\right) \lambda e^{-\lambda s} e^{-\alpha_{nt}(s)T} ds + N\int_{s_0}^{\infty} \lambda e^{-\lambda s} e^{-\alpha_{nt}(s)T} ds \quad (13.26)$$

10. Finally, the total mortality is given by

$$M(s_0) = N - \left[F\int_{0}^{s_0} \lambda e^{-\lambda s} e^{-\alpha_{t}(s)T} ds + N\int_{0}^{s_0} \left(1-\frac{F}{N}\right) \lambda e^{-\lambda s} e^{-\alpha_{nt}(s)T} ds + N\int_{s_0}^{\infty} \lambda e^{-\lambda s} e^{-\alpha_{nt}(s)T} ds\right] \quad (13.27)$$

11. Now, to calculate the optimal transplantation strategy, we determine the MELD score that can be transplanted and find either s such that $\min[M(s)]$ or s such that $M(s) = M(s_M)$. The result can be seen in Fig. 13.7.

Note that Eq. (13.26) has no optimum MELD score that minimizes the total mortality. In fact, because of the higher mortality observed in this cohort of transplanted patients than nontransplanted, the MELD score at presentation that minimize the total mortality is the maximum possible value of score, that is, it is better, according to this analysis, not to transplant. Further tests, however, are still necessary to support this conclusion.

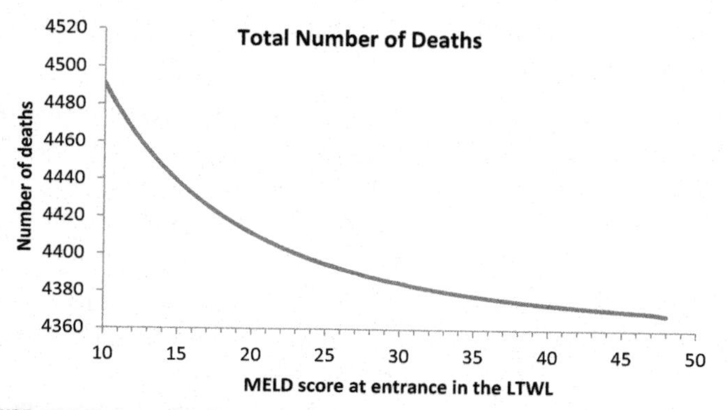

■ **FIGURE 13.7** Total mortality for transplanted and nontransplanted patients as a function of MELD score at presentation.

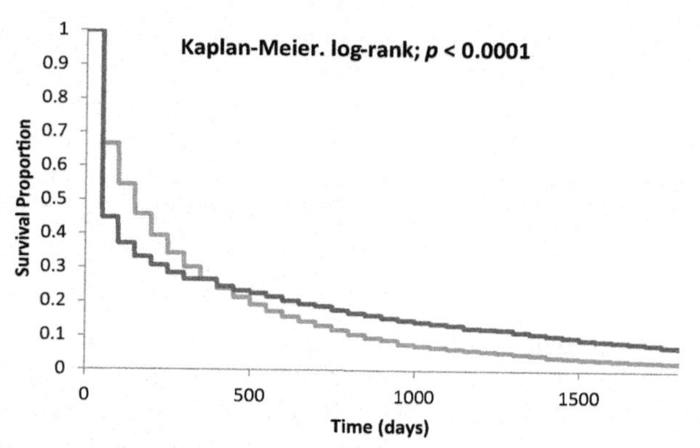

■ **FIGURE 13.8** Kaplan–Meier survival curve for transplanted (red line *[black in print version]*) and nontransplanted (blue line *[dark gray in print version]*) patients.

In Fig. 13.8, we show the survival curve of both transplanted and nontransplanted patients for the Kaplan–Meier analysis.

Note that the mortality is greater for transplanted patients in the first year after the transplantation, leveling off from thereafter. As mentioned above, however, further analysis will clarify what is happening with the transplantation program in the State of São Paulo.

13.2 THE FUTURE OF LIVER TRANSPLANTATION

Bioengineering approaches

The liver is one of the most complex organs owing to its highly complex and interwoven vascular, lymph, and biliary network. Thus, both the cellular and biochemical composition, as well as the structural organization of the liver, is crucial for the development of fully functional liver substitutes (Mazza et al., 2017).

One approach toward developing a partial liver graft is the fabrication of cell-laden constructs in vitro using molding or 3D printing. Successful culture of hepatocytes and hepatocytelike cells has been shown in tissue models comprised of synthetic materials (Mohanty et al., 2016), as well as natural materials (Lee et al., 2016, 2017).

3D printing, hydrogel-based tissue fabrication, and the use of native decellularized liver extracellular matrix as a scaffold are used to develop whole or partial liver substitutes. The current focus is on developing a functional liver graft through achieving a nonleaky endothelium and a fully constructed bile duct. Use of cell therapy as a treatment is less invasive and less costly compared with transplantation; however, lack of readily available cell sources with low or no immunogenicity and contradicting outcomes of clinical trials are yet to be overcome. Liver bioengineering is advancing rapidly through the development of in vitro and in vivo tissue and organ models. Although there are major challenges to overcome, through optimization of the current methods and successful integration of induced pluripotent stem cells, the development of readily available, patient-specific liver substitutes can be achieved.

Current applications of liver bioengineering include fabricating partial liver grafts through 3D bioprinting and using hydrogel-based tissues for repairing the lost liver function (a). Recently, the focus has been shifted toward optimizing the decellularization of donors livers rejected for transplantation and recellularization with induced pluripotent stem cells—derived hepatocytes to render them eligible for transplantation (b). This is illustrated by Fig. 13.9 below.

Artificial organs

Bioartificial liver is a novel device that has shown to reduce the severity of liver disease and improve survival in laboratory testing on pigs (Nyberg, 2019). Developed by the Mayo Clinic, the new device, called Mayo Spheroid Reservoir Bioartificial Liver, has demonstrated ammonia clearance a higher level that the reports from earlier generations of artificial livers (Fig. 13.10).

At this stage of development, the bioartificial liver is indicated to patients who have acute liver failure while they are waiting for a transplant, for patients with overdose of medication that caused liver failures but have good

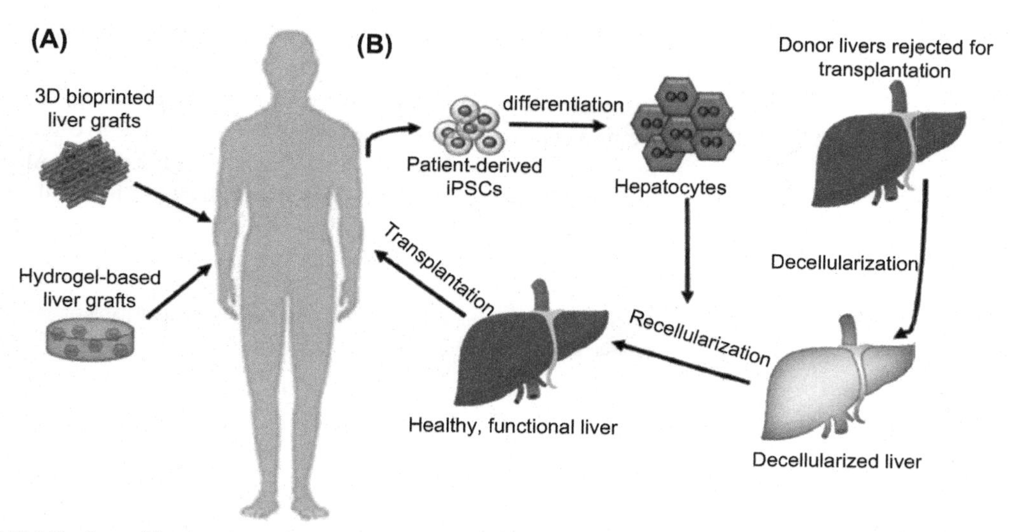

■ **FIGURE 13.9** The future of liver bioengineering. *From Acun, A., Oganesyan, R., Uygun, B.E., 2019. Liver bioengineering: promise, pitfalls, and hurdles to overcome. Curr. Transplant. Rep. https://doi.org/10.1007/s40472-019-00236-3.*

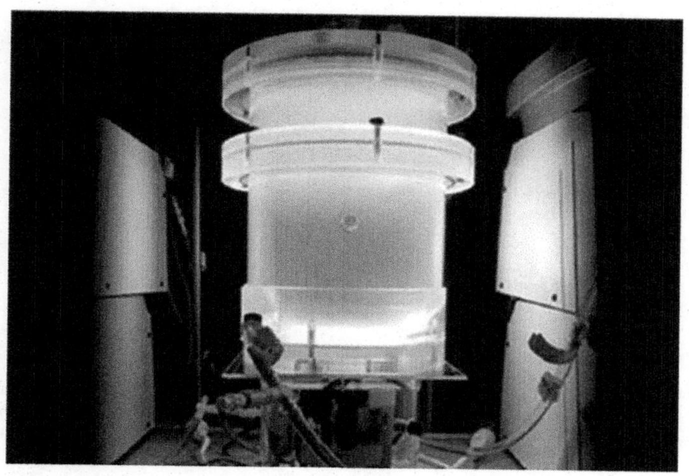

■ **FIGURE 13.10** The Mayo Spheroid Reservoir Bioartificial Liver. *From Nyberg, 2019. Used with permission of Mayo Foundation for Medical Education and Research. All rights reserved.*

perspective of regeneration and for patients with liver failure but are not candidates for transplantation due to other health concerns.

Developed with the same objective as hemodialysis machines, the bioartificial liver equipment works from outside the patient's body. It differ from the dialysis machines, which are entirely artificial; the bioartificial liver equipment is a hybrid extracorporeal device in that it contains hepatocytes of porcine or human origin as a biological source of liver function. As such, provided that it can be perfected to be more feasible of permanent use, it can in the future evolve to be a technology to substitute liver transplant. At the current stage, however, it aims to serve as a bridge to transplant and can be used up to a maximum of 2 weeks. Anyhow, it certainly represents a new paradigm in the treatment of liver failure.

REFERENCE

Nyberg SL. Available at: https://www.mayo.edu/research/labs/artificial-liver-transplantation/projects/spheroid-reservoir-bioartificial-liver. Assessed 10 July 2019.

References

Abt, P., Crawford, M., Desai, N., Markmann, J., Olyhoff, K., Shaked, A., 2003. Liver transplantation from controlled non-heart-beating donors: an increased incidence of biliary complications. Transplantation 75, 1659.

Abt, P.L., Desai, N.M., Crawford, M.D., 2004. Survival following liver transplantation from non-heart-beating donors. Ann. Surg. 239, 87–92.

Acun, A., Oganesyan, R., Uygun, B.E., 2019. Liver bioengineering: promise, pitfalls, and hurdles to overcome. Curr. Transplant. Rep. Available at: https://doi.org/10.1007/s40472-019-00236-3.

Ahmad, J., Downey, K.K., Akoad, M., Cacciarelli, T.V., 2007. Impact of the MELD score on waiting time and disease severity in liver transplantation in United States veterans. Liver Transplant. 13 (11), 1564–1569.

Alvarez-Rodrigues, J., Barrio-Yesa, R., Torrente-Sierra, J., 1995. Post-transplant long-term outcome of kidneys obtained from asystolic donors maintained under extracorporeal cardiopulmonary bypass. Transplant. Proc. 27, 2903–2905.

Amaku, M., Coutinho, F.A., Chaib, E., Massad, E., 2013. The impact of hepatitis A virus infection on hepatitis C virus infection: a competitive exclusion hypothesis. Bull. Math. Biol. 75 (1), 82–93.

Amaku, M., Burattini, M.N., Chaib, E., Coutinho, F.A.B., Greenhalgh, D., Lopez, L.F., Massad, E., 2017. Estimating the prevalence of infectious diseases from under-reported age-dependent compulsorily notification databases. Theor. Biol. Med. Model. 14 (1), 23.

Austin, M.T., Poulose, B.K., Ray, W.A., Arbogast, P.G., Feurer, I.D., Pinson, C.W., 2007. Model for End-Stage Liver Disease: did the new liver allocation policy affect waiting list mortality? Arch. Surg. 142, 1079–1085.

Azeka, E., Auler Júnior, J.O., Fernandes, P.M., et al., 2009. Regsitry of Hospital das Clinicas of the University of São Paulo Medical School: first official solid organ and tissue transplantation report-2008. Clinics 64, 127.

Backman, L., Appelkvist, E.L., Ringden, O., Dallner, G., 1988. Glutathione transferase in the urine: a marker for post-transplant tubular lesions. Kidney Int. 33, 571–577.

Bains, J.C., Sandford, R.M., Brook, N.R., Hosgood, S.A., Lewis, G.R.R., Nicholson, M.L., 2005. Comparison of renal allograft fibrosis after transplantation from heart-beating and non-heart-beating donors. Br. J. Surg. 92, 113–118.

Baiochi, E., Nelson, L.C., Colas, O.R., 2007. Blood types group frequency and ABO/RH incompatibility in pregnant and new-born. Rev. Assoc. Med. Bras. 53, 44.

Balupuri, S., Buckley, P., Mohamad, M., Chidambaram, V., Gerstenkorn, C., Sen, B., et al., 2000a. Early results of anon-heratbeating donor (NHBD) programme with machine perfusion. Transpl. Int. 13 (1), S255–S258.

Balupuri, S., Buckley, P., Snowden, C., 2000b. The troble with kidneys derived from non-heart-beating donor: a single center 10-year experience. Transplantation 15 (69), 842.

Balupuri, S., Mantle, D., Mohamed, M., Shenton, B., Gok, M., Soomro, N., et al., 2001. Machine perfusion and viability assessment of Non-Heart-Beating Donor kidney – a single centre result. Transplant. Proc. 33, 1119–1120.

Baxby, K., Taylor, R.M., Anderson, M., Johnson, R.W., Swinney, J., 1974. Assessment of cadaveric kidneys for transplantation. Lancet 2, 977.

Beecher, H.K., 1968. A definition of irreversible coma. Report of the ad hoc committee of the Harvard Medical School to examine the definition of brain death. JAMA 205, 337.

Biasi, F., Poli, G., Salizzoni, M., Cerutti, E., Battista, S., Mengozzi, G., 2002. Effect of perioperative infusion of antioxidants on neutrophil activation during liver transplantation in humans. Transplant. Proc. 34, 755.

Biggins, S.W., Kim, W.R., Terrault, N.A., Saab, S., Balan, V., Schiano, T., Benson, J., Therneau, T., Kremers, W., Wiesner, R., Kamath, P., Klintmalm, G., 2006. Evidence based incorporation of serum sodium concentration into MELD. Gastroenterology 130, 1652−1660.

Billingham, R.E., Trent, L., Medawar, P.B., 1953. Actively acquired tolerance of foreign cells. Nature 172, 176−199.

Bismuth, H., Chiche, L., Adam, R., Castaing, D., Diamond, T., Dennison, A., 1993. Liver resection versus transplantation for hepatocellular carcinoma in cirrhotic patients. Ann. Surg. 10, 145−151.

Bismuth, H., Majno, P.E., Adam, R., 1999. Liver transplantation for hepatocellular carcinoma. Semin. Liver Dis. 19, 311−322.

Boin, I.F., Pracucho, E.M., Rique, M.C., Reno, R.R., Robertoni, D.B., Silva, P.V., Rosim, E.T., Soares, A.B., Escanhoela, C.A., Leonardi, M.I., Souza, J.R., Leonardi, L.S., 2008. Expanded Milan criteria on pathological examination after liver transplantation: analysis of preoperative data. Transplant. Proc. 40, 777−779.

Brandão, S.L., Fuchs, A.L., Gleisner, et al., 2009. MELD and other predictors of survival after liver transplantation. Clin. Transplant. 23 (2), 220−227.

Briceno, J., Padillo, J., Rufián, S., Solórzano, G., Pera, C., 2005. Assignment of steatotic livers by the mayo model for end-stage liver disease. Transpl. Int. 18 (5), 577−583.

Brook, N.R., Waller, J.R., Nicholson, M.L., 2003a. Nonheart-beating kidney donation: current practice and future developments. Kidney Int. 63, 1516.

Brook, N.R., White, S.A., Waller, J.R., Veitch, P.S., Nicholson, M.L., 2003b. Non-heart beating donor kidneys with delayed graft function have a superior graft survival compared with conventional heart-beating donor kidneys that developed delayed graft function. Am. J. Transplant. 3, 614−618.

Brook, N.A., Walter, J.R., Richardson, A.C., Bradley, J.A., Andrews, P.A., Koffman, G., et al., 2004. A report on the activity and clinical outcomes of renal non-heart-beating donor transplantation in the United Kingdom. Clin. Transplant. 18, 627−633.

Bruix, J., Fuster, J., Llovet, J.M., 2003. Liver transplantation for hepatocellular carcinoma: Foucault pendulum versus evidence-based decision. Liver Transpl. 9, 700−702.

Bruix, J., Sherman, N., Llovet, J.M., Beaugrand, M., Lencioni, R., Burroughs, A.K., Christensen, E., Pagliaro, L., Colombo, M., Rodés, J., EASL panel of Experts on HCC, 2005. Clinical management on hepatocellular carcinoma. Conclusions of the Barcelona-2000 EASL Conference. J. Hepatol. 35, 421−430.

Bulmer, M.G., 1985. The Mathematical Theory of Quantitative Genetics. Clarendon Press, Oxford, England.

Burdick, J.F., Rosendale, J.D., McBride, M.A., Kauffman, M., Bennet, L.E., 1997. National impact of pulsatile perfusion on cadaveric kidney transplantation. Transplantation 64, 1730.

Burrel, M., Llovet, J.M., Ayuso, C., Iglesias, C., Sala, M., Miquel, R., Caralt, T., Ayuso, J.R., Solé, M., Sanchez, M., Brú, C., Bruix, J., Barcelona Clínic Liver Cancer Group, 2003. MRI angiography is superior to helicoidal CT for detection of HCC prior liver transplantation: an explants correlation. Hepatology 38, 1034−1042.

Busuttil, R.W., Klintmalm, G.B.G., 2015. Transplantation of the Liver. Elsevier, Philadelphia.

Busuttil, R.W., Farmer, D.G., Yersiz, H., Hiatt, J.R., McDiarmid, S.V., Goldstein, L.I., Saab, S., Han, S., Durazo, F., Weaver, M., Cao, C., Chen, T., Lipshutz, G.S., Holt, C., Gordon, S., Gornbein, J., Amersi, F., Ghobrial, R.M., 2005. Analysis of long-term outcomes of 3200 liver transplantations over two decades: a single center experience. Ann. Surg. 241, 905−916.

Butterworth, P.C., Taub, N., Doughman, T.M., Horsburgh, T., Veitch, P.S., Bell, P.R.F., et al., 1997. Are kidneys from non-heart-beating donors second class organs? Transplant. Proc. 29, 3567−3568.

Calne, R.Y., 1960. The rejection of renal allograft. Inhibition in dogs by 6-mercaptopurine. Lancet 1, 417−418.

Calne, R.Y., 1983. Liver Transplantation. Grune & Stratton, New York.

Calne, R.Y., Willians, R., 1979. Liver transplantation. Curr. Probl. Surg. 16, 3−44.

Calne, R.Y., Rolles, K., White, D.J.G., et al., 1979. Cyclosporine A initially as the only immunossupressant in 34 recipients of cadaveric organs: 32 kidneys, 2 pancreases, and 2 livers. Lancet 2, 1033–1036.

Campos Freire, J.G., Sabbaga, E., Cabral, A.D., Verginelli, G., Goes, G.M., Ianhez, L.E., 1968. Renal homotransplantation. Analysis of the first 15 cases at university of São Paulo School of medicine. Rev. Assoc. Med. Bras. 14, 133.

Casavilla, A., Ramirez, C., Shapiro, R., Nghiem, D., Miracle, K., Fung, J.J., et al., 1995. Experience with liver and kidney allografts from non-heart-beating donors. Transplantation 59, 197.

Castaneda, M.P., Swatecka-Urban, A., Mitsnefes, M.M., Feuerstein, D., Kaskel, F.J., Tellis, V., 2003. Activation of mitochondrial apoptotic pathways in human renal allografts after ischemia reperfusion injury. Transplantation 76, 50.

Castelao, A.M., Grino, J.M., Gonzalez, C., 1993. Update of our experience in long-term renal function of kidneys transplanted from non-heart-beating cadáver donors. Transplant. Proc. 25, 1513.

Chaib, E., 2008. Non-heart-beating donors in England. Clinics 63, 121.

Chaib, E., Massad, E., 2005. Liver transplantation: waiting list dynamics in the state of Sao Paulo, Brazil. Transplant. Proc. 37, 4329.

Chaib, E., Massad, E., 2007. Comparing the dynamics of kidney and liver transplantation waiting list in the state of Sao Paulo, Brazil. Transplantation 84, 1209–1211.

Chaib, E., Massad, E., 2008a. The potential impact of using donations after cardiac death (DCD) on liver transplantation program and waiting list in the state of Sao Paulo, Brazil. Liver Transplant. 14, 1732.

Chaib, E., Massad, E., 2008b. Expected number of deaths in the liver transplantation waiting list in the state of São Paulo, Brazil. Transpl. Int. 21, 290–291.

Chaib, E., Massad, E., 2008c. Calculating the liver lobe weight for transplantation. Transpl. Int. 10, 704–706.

Chaib, E., Coimbra, B.G., Galvão, F.H., Tatebe, E.R., Shinzato, M.S., D'Albuquerque, L.A., Massad, E., 2012. Does anti-hepatitis B virus vaccine make any difference in long-term number of liver transplantation? Clin. Transplant. 10, E590–E595.

Chaib, E., Molinari, H., Morales, M.M., Bordalo, M., Raia, S., 1994. Estado atual do "split-liver" como opção técnica para o transplante de fígado. Revisão da literatura. Rev. Hosp. Clin. Fac. Med. Sao Paulo 49, 53–56.

Chaib, E., Antonio, L.G.M., Ishida, R.Y., Feijó, L.F.A., Morales, M.M., Rodrigues, M.B., et al., 1995. Estudo do sistema venoso hepático e sua aplicação na técnica de transplante de fígado chamada "split-liver". Rev. Hosp. Clin. Fac. Med. Sao Paulo 50, 49–51.

Chaib, E., Bertevello, P., Saad, W.A., et al., 2007. The main hepatic anatomic variations for the purpose of split-liver transplantation. Hepatogastroenterology 54, 688–692.

Chaib, E., de Oliveira, M.C., Galvão, F.H., Silva, F.D., D'Albuquerque, L.A., Massad, E., 2010. Theoretical impact of an anti-HCV vaccine on the annual number of liver transplantation. Med. Hypotheses 75, 324.

Chaib, E., Figueira, E.R.R., Brunheroto, A., Gatti, A.P., Fernandes, D.V., D'Albuquerque, L.A.C., 2013a. Does the patient selection with MELD score improve short-term survival in liver transplantation? Arq. Br. Cir. Dig. 26 (4), 324–327.

Chaib, E., Amaku, M., Coutinho, F.A., Lopez, L.F., Burattini, M.N., D'Albuquerque, L.A., Massad, E., 2013b. A mathematical model for optimizing the indications of liver transplantation in patients with hepatocellular carcinoma. Theor. Biol. Med. Model. 10, 60.

Chawla, Y.K., Kashinath, R.C., Duseja, A., Dhiman, R.K., 2011. Predicting mortality across a broad spectrum of liver disease—an assessment of Model for End-Stage Liver Disease (MELD), Child–Turcotte–Pugh (CTP), and creatinine-modified CTP scores. J. Clin. Exp. Hep. 10, 161–168.

Cillo, U., Vitale, A., Bassanello, M., Boccagni, P., Brolese, A., Zanus, G., Burra, P., Fagiuoli, S., Farinati, F., Rugge, M., D'Amico, D.F., 2004. Liver transplantation for the treatment of moderately or well-differentiation hepatocellular carcinoma. Ann. Surg. 239, 150–159.

CHMSEDBD, 1968. A definition of irreversible coma. Report of the ad hoc committee of the Harvard Medical School to examine the definition of brain death. JAMA 205, 337–340.

Cho, Y.M., Terasaki, P.I., Cheka, J.M., Gjertson, D.W., 1998. Transplantation from donors whose hearts have stopped beating. N. Engl. J. Med. 338, 221.

Christensen, E., Schlichting, P., Fauerholdt, L., et al., 1984. Prognostic value of Child-Turcotte criteria in medically treated cirrhosis. Hepatology 4, 430–435.

Crespo, G., Mariño, Z., Navasa, M., Forns, X., 2012. Viral hepatitis in liver transplantation. Gastroenterology 142, 1373–1383 e1. https://doi.org/10.1053/j.gastro.2012.02.011.

Cuomo, O., Perrella, A., Arenga, G., 2008. Model for End-Stage Liver Disease (MELD) score system to evaluate patients with viral hepatitis on the waiting list: better than the Child-Turcotte-Pugh (CTP) system? Transplant. Proc. 40, 1906–1909.

D'Alessandro, A.M., Hoffmann, R.M., Knechtle, S.J., et al., 2000. Liver transplantation from controlled non-heart-beating donors. Surgery 128, 579.

Daemen, J.W., Kootstra, G., Wijnen, R.M., Yin, M., Heineman, E., 1994. Nonheart-beating donors: the Maastricht experience. Clin. Transpl. 7, 303–306.

Daemen, J.W., de Wit, R.J., Bronkhorst, M.W., Yin, M., Heineman, E., Kootstra, G., 1996. Non-heart-beating donor program contributes 40% of kidneys for transplantation. Transplant. Proc. 28, 105.

Daemen, J.H., de Vries, B., Oomen, A.P., Demeester, J., Kootstra, G., 1997. Effect of machine perfusion preservation on delayed graft function in non-heart-beating donor kidneys-early results. Transpl. Int. 10, 317.

Daemen, J.H.C., Oomen, A.P.A., Belgers, E.H.J., 1997. Procurement and transplantation of kidneys from non-heart-beating donors – an overview. Dig. Surg. 14, 333.

Daemen, J.W., Oomen, A.P.A., Janssen, M.A., Schoot, L.V.D., Kreel, B.K., Heineman, E., et al., 1997. Glutathione S transferase as predictor of functional outcome in transplantation of machine preserved non-heartbeating donor kidneys. Transplantation 63, 89–93.

DATASUS. Available at: http://tabnet.datasus.gov.br/cgi/deftohtm.exe?sim/cnv/obtsp.def. [Accessed February 2008].

de Carvalho, H.B., Mesquita, F., Massad, E., et al., 1996. HIV and infections of similar transmission patterns in a drug injectors community of Santos, Brazil. J. Acquir. Immune Defic. Syndr. Hum. Retrovirol. 12, 84.

Decaens, T., Roudot-Thoraval, F., Hadni-Bresson, S., Meyer, C., Gugenheim, J., Durand, F., Bernard, P.H., Boillot, O., Sulpice, L., Calmus, Y., Hardwigsen, J., Ducerf, C., Pageaux, G.P., Dharancy, S., Chazouilleres, O., Cherqui, D., Duvoux, C., 2006. Impact of UCSF criteria according to pre- and post-OLT tumor features: analysis of 479 patients listed for HCC with a short waiting time. Liver Transplant. 12, 1761–1769.

Demertzis, S., Langer, F., Graeter, T., Dwenger, A., Georg, T., Schafers, H.J., 1999. Amelioration of lung reperfusion injury by L- and E-selectin blockade. Eur. J. Cardiothor. Surg. 16, 174.

Duffy, J.P., Vardanian, A., Benjamin, E., Watson, M., Farmer, D.G., Ghobrial, R.M., Lipshutz, G., Yersiz, H., Lu, D.S., Lassman, C., Tong, M.J., Hiatt, J.R., Busuttil, R.W., 2007. Liver transplantation criteria for hepatocellular carcinoma should be expanded: a 22-year experience with 467 patients at UCLA. Ann. Surg. 246, 502–509.

Dunlop, P., Varty, K., Veitch, P.S., 1995. Non-heart-beating donors: the Leicester experience. Transplant. Proc. 27, 2940.

Durand, F., Belguiti, J., 2003. Liver transplantation for hepatocellular carcinoma: should we push the limits? Liver Transplant. 9, 697–699.

Facciuto, M., Heidt, D., Guarrera, J., Bodian, C.A., Miller, C.M., Emre, S., Guy, S.R., Fishbein, T.M., Schwartz, M.E., Sheiner, P.A., 2000. Retransplantation for late liver graft failure: predictors of mortality. Liver Transplant. 6, 174–179.

Fan, J., Yang, G.S., Fu, Z.R., Peng, Z.H., Xia, Q., Peng, C.H., Qian, J.M., Zhou, J., Xu, Y., Qiu, S.J., Zhong, L., Zhou, G.W., Zhang, J.J., 2009. Liver transplantation outcomes in 1,078 hepatocellular carcinoma patients: a multi-center experience in Shanghai, China. J. Cancer Res. Clin. Oncol. 135, 1403–1412.

Feldman, H.I., Gaynaer, R., Berlin, J.R., 1996. Delayed function reduces renal allograft survival independent of acute rejection. Nephrol. Dial. Transplant. 11, 1306–1313.

Figueras, J., Jaurrieta, E., Valls, C., Benasco, C., Rafecas, A., Xiol, X., Fabregat, J., Casanovas, T., Torras, J., Baliellas, C., Ibañez, L., Moreno, P., Casais, L., 1997. Survival after liver transplantation in cirrhotic patients with or without hepatocellular carcinoma: a comparative study. Hepatology 25, 1485–1489.

Figueras, J., Jaurrieta, E., Valls, C., Ramos, E., Serrano, T., Rafecas, A., Fabregat, J., Torras, J., 2000. Resection or transplantation for hepatocellular carcinoma in cirrhotic patients: outcomes based on indicated treatment strategy. J. Am. Coll. Surg. 10, 580–587.

Fink, M.A., Angus, P.W., Gow, P.J., et al., 2005. Liver transplant recipient selection: MELD vs. clinical judgment. Liver Transplant. 11 (6), 621–626.

Freeman Jr., R.B., Wiesner, R.H., Edwards, E., Harper, A., Merion, R., Wolfe, R., 2004a. Results of the first year of the new liver allocation plan. Liver Transplant. 10 (1), 7–15.

Freeman Jr., R.B., Wiesner, R.H., Roberts, J.P., McDiarmid, S., Dykstra, D.M., Merion, R.M., 2004b. Improving liver allocation: MELD and PELD. Am. J. Transplant. 4 (9), 114–131.

Gallegos-Orozco, J.F., Yosephy, A., Noble, B., Aqel, B.A., Byrne, T.J., Carey, E.J., et al., 2009. Natural history of post-liver transplantation hepatitis C: a review of factors that may influence its course. Liver Transplant. 15 (12), 1872–1881.

Garcia, V.D., 2005. Increase number of brazilian transplantations. Have we got anything to celebrate? J. Reg. São Paulo Council Med. 30, 1.

Garcia-Rinaldi, R., Lefrak, E.A., Defore, W.W., 1975. In situ preservation of cadáver kidneys for transplantation: laboratory observations and clinical application. Ann. Surg. 182, 576–584.

Garcia-Valdecasas, J.C., Tabet, J., Valero, R., et al., 1999. Evaluation of ischemic injury during liver procurement from non-heart-beating donors. Eur. Surg. Res. 31, 447.

Gershenfeld, N.A., 1999. The Nature of Mathematical Modeling. Cambridge. University Press.

Gerstenkorn, C., Deardon, D., Koffman, C.G., Papalois, V.E., Andrews, P.A., 2002. Outcome of renal allografts from non-heart-beating donors with delayed graft function. Transpl. Int. 15, 660–663.

Gok, M.A., Shenton, B.K., Buckley, P.E., Balupuri, S., Soomro, N., Manas, D.M., et al., 2002a. Long-term renal function in kidneys from non-heart-beating donors: a single-center experience. Transplant. Proc. 34, 2598–2599.

Gok, M.A., Buckley, P.E., Shenton, B.K., Balupuri, S., El-Sheik, M.A.F., Robertson, H., et al., 2002b. Long-term renal function in kidneys from non-heart-beating donors: a single-center experience. Transplantation 74, 664–669.

Gok, M.A., Shenton, B.K., Buckley, P.E., Peaston, R., Cornell, C., Soomro, N., et al., 2003. How to improve the quality of kidneys from non-heart-beating donors: a randomized controlled trial of thrombolysis in non-heartbeating donors. Transplantation 76, 1714–1719.

Gok, M.A., Asher, J.F., Shenton, B.K., Rix, D., Soomro, N.A., Jaques, B.C., et al., 2004a. Graft function after kidney transplantation from non-heart-beating donors according toaastricht category. J. Urol. 172, 2331–2334.

Gok, M.A., Gupta, A., Olschewski, P., Bhatti, A., Shenton, B.K., Robertson, H., et al., 2004b. Renal transplants from non-heart-beating paracetamol overdose donors. Clin. Transplant. 18, 541–546.

Gok, M.A., Shenton, B.K., Pelsers, M., Whitwood, A., Mantle, D., Cornell, C., et al., 2006. Ischemia-reperfusion injury in cadaveric nonheart beating, cadaveric heart beating and live donor renal transplants. J. Urol. 175, 641–647.

Gondolesi, G., Muñoz, L., Matsumoto, C., Fishbein, T.M., Sheiner, P., Emre, S., Miller, C., Schwartz, M.E., 2002. Hepatocellular carcinoma: a prime indication for living donor liver transplantation. J. Gastrointest. Surg. 6, 102–107.

Gondolesi, G.E., Roayaie, S., Munoz, L., Kim-Schluger, L., Schiana, T., Fishbein, T.M., Emre, S., Miller, C.M., Schwartz, M.E., 2004. Adult living donor liver transplantation for patients with hepatocellular carcinoma extending UNOS priority criteria. Ann. Surg. 239, 142–149.

Gonzalez-Segura, C., Castelao, A.M., Torras, J., 1998. A good alternative to reduce the kidney shortage – kidneys from non-heart-beating donors. Transplantation 65, 1465–1470.

Gruessner, R.W., Benedetti, E. (Eds.), 2008. Living Donor Organ Transplantation. McGraw-Hill Medical, New York, NY.

Gruttadauria, S., Vizzini, G., Biondo, D., Mandalà, L., Volpes, R., Palazzo, U., Gridelli, B., 2008. Critical use of extended criteria donor liver grafts in adult-to-adult whole liver transplantation: a single-center experience. Liver Transplant. 14, 220–227.

Guillard, G., Rat, P., Haas, O., Letourneau, B., Isnardon, J.P., Favre, J.P., 1993. Renal harvesting after in situ cooling by intra-aortic double balloon catheter. Transplant. Proc. 25, 1505.

Haisch, C., Green, E., Brasile, L., 1997. Predictors of graft outcome in warm ischemically damaged organs. Transplant. Proc. 29, 3424.

Hajarizadeh, B., Grebely, J., Dore, G.J., 2013. Epidemiology and natural history of HCV infection. Nat. Rev. Gastroenterol. Hepatol. 10, 553–562.

Heineman, E., Daemen, J.H.C., Kootstra, G., 1995. Non heart-beating donors: methods and techniques. Transplant. Proc. 29, 2895–2897.

Hepatitis B Foundation UK, 2007. Rising curve, chronic HBV infection in the UK. In: Chronic Hepatitis B Prevalence Form. Hepatitis B Foundation UK.

Herrero, J.I., Sangro, B., Quiroga, J., Pardo, F., Herraiz, M., Cienfuegos, J.A., Pietro, J., 2001. Influence of tumor characteristics on the outcome of liver transplantation among patients with liver cirrhosis and hepatocellular carcinoma. Liver Transplant. 7, 631–636.

Herrero, J.I., Sangro, B., Pardo, F., Quiroga, J., Iñarrairaegui, M., Rotellar, F., Montiel, C., Alegre, F., Prieto, J., 2008. Liver transplantation in patients with hepatocellular carcinoma across Milan criteria. Liver Transplant. 14, 272–278.

Hilleman, M.R., 1993. Plasma derived hepatitis B vaccine: a breakthrough in preventive medicine. In: Ellis, R. (Ed.), Hepatitis B Vaccines in Clinical Practice. Marcel Dekker, New York, p. 17.

Hilleman, M.R., 2011. Three decades of hepatitis vaccinology in historic perspective. A paradigm of successful pursuits. In: Plotkin, S.A. (Ed.), History of Vaccine Development. Springer Verlag, New York, p. 233.

Hoel, P.G., 1984. Introduction to Mathematical Statistics. John Wiley & Sons, New York.

Hoshinaga, K., Fujita, T., Naide, Y., 1995. Early prognosis of 263 renal allografts harvested from non-heart-beating cadavers using an in situ cooling technique. Transplant. Proc. 27, 703.

Iwasaki, Y., Fukao, K., Iwasaki, H., 1990. Moral principles of kidney donation in Japan. Transplant. Proc. 22, 963.

Iwatsuki, S., Starzl, T.E., Sheahan, D.G., Yokoyama, I., Demetris, A.J., Todo, S., Tzakis, A.G., Van Thiel, D.H., Carr, B., Selby, R., et al., 1991a. Hepatic resection versus transplantation for hepatocellular carcinoma. Ann. Surg. 214, 221–228.

Iwatsuki, S., Starzl, T.E., Sheahan, D.G., Yokoyama, I., Demetris, A.J., Todo, S., Tzakis, A.G., Van Thiel, D.H., Carr, B., Selby, R., 1991b. Hepatic resection versus transplantation for hepatocellular carcinoma. Ann. Surg. 10, 221–229.

Jonas, S., Bechstein, W.O., Steinmüller, T., Herrmann, M., Radke, C., Berg, T., Settmacher, U., Neuhaus, P., 2001. Vascular invasion and histhopathologic grading determine outcome after liver transplantation for hepatocellular carcinoma in cirrhosis. Hepatology 33, 1080–1086.

Kamath, P.S., Wiesner, R.H., Malinchoc, M., et al., 2000. A model to predict survival in patients with end-stage liver disease. Hepatology 33 (2), 464–470.

Kim, R.D., Nazarey, P., Katz, E., Chari, R.S., 2004. Laparoscopic staging and tumor ablation for hepatocellular carcinoma in Child C cirrhotics evaluated for orthotopic liver transplantation. Surg. Endosc. 18, 39–44.

Kim, W.R., Biggins, S.W., Kremers, W.K., et al., 2008. Hyponatremia and mortality among patients on the liver transplant wait list. N. Engl. J. Med. 359, 1018–1026.

Kofmann, C.G., Bewick, M., Chang, R.W.S., et al., 1993. Comparative study of the use of systolic and asystolic kidney donors between 1988 and 1991. Transplant. Proc. 25, 1527.

Kootstra, G., Wijnen, R., van Hooff, J.P., Vad der Linden, C.J., 1991. Twenty percent more kidneys through a nonheart beating program. Transplant. Proc. 23, 910–911.

Kootstra, G., Daemen, J.H.C., Oomen, A.P.A., 1995a. Categories of non-heartbeating donors. Transplant. Proc. 27, 2893–2894.

Kootstra, G., Arnold, R.M., Bos, M.A., 1995b. Round table discussion on non-heart-beating donors. Transplant. Proc. 27, 2935.

Kootstra, G., Kievit, J.K., Heineman, E., 1997. The non-heart beating donor. Br. Med. Bull. 53, 844.

Kwiatkosk, A., Danielewicz, R., Polak, W., Michalak, G., Paczek, Walaszewski, J., et al., 1996. Storage by continuous hypothermic perfusion for kidney harvested from hemodynamically unstable donors. Transplant. Proc. 28, 306.

Laskowski, I.A., Pratschke, J., Wilhelm, M.J., et al., 1999. Non-heartbeating kidney donors. Clin. Transpl. 13, 281.

Lawrence, S.P., 2000. Advances in the treatment of hepatitis C. Adv. Intern. Med. 45, 65–105.

Lee, S.G., Hwang, S., Moon, D.B., Ahn, C.S., Kim, K.H., Sung, K.B., Ko, G.Y., Park, K.M., Ha, T.Y., Song, G.W., 2008. Expanded indication criteria of living donor liver transplantation for hepatocellular carcinoma at one large-volume center. Liver Transplant. 14, 935–945.

Lee, J.W., Choi, Y.J., Yong, W.J., Pati, F., Shim, J.H., Kang, K.S., et al., 2016. Development of a 3D cell printed construct considering angiogenesis for liver tissue engineering. Biofabrication 8, 015007.

Lee, H., Han, W., Kim, H., Ha, D.H., Jang, J., Kim, B.S., et al., 2017. Development of liver decellularized extracellular matrix bioink for threedimensional cell printing-based liver tissue engineering. Biomacromolecules 18, 1229–1237.

Li, J., Yan, L.N., Yang, J., Chen, Z.Y., Li, B., Zeng, Y., Wen, T.F., Zhao, J.C., Wang, W.T., Yang, J.Y., Xu, M.Q., Ma, Y.K., 2009. Indicators of prognosis after liver transplantation in Chinese hepatocellular carcinoma patients. World J. Gastroenterol. 15, 4170–4176.

Light, J.A., Kowalski, A.E., Ritchie, W.O., 1996. New profile of cadaveric donors: what are the kidney donor limits? Transplant. Proc. 28, 17.

Light, J.A., Sasaki, T.M., Aquino, A.O., Barhyte, D.Y., Gage, F., 2000. Excellent long-term graft survival with kidneys from the uncontrolled non-heartbeating donor. Transplant. Proc. 32, 186.

Lindskog, P., Ljung, L., May 1997. Ensuring certain physical properties in black box models by applying fuzzy techniques. IFAC Proc. Vol. 30 (11).

Ljung, L., 1987. System Identification: Theory for the User. Prentice-Hall, Englewood Cliffs, NJ.

Llovet, J.M., Bruix, J., Fuster, J., Castells, A., Garcia-Valdecasas, J.C., Grande, L., Franca, A., Brú, C., Navasa, M., Ayuso, M.C., Solé, M., Real, M.I., Vilana, R., Rimola, A., Visa, J., Rodés, J., 1998. Liver transplantation for small hepatocellular carcinoma: the tumor- node-metastatsis classification does not have prognostic power. Hepatology 27, 1572–1577.

Llovet, J.M., Bustamenate, J., Castells, A., Vilana, R., Ayuso, M. d C., Sala, M., 1999a. Natural hystory of untreated nonsurgical hepatocellular carcinoma: rationale for the design and evaluation of therapeutics trials. Hepatology 29, 62–67.

Llovet, J.M., Fuster, J., Bruix, J., 1999b. Intention-to-treat analysis of surgical treatment for early hepatocellular carcinoma: resection versus transplantation. Hepatology 30, 1434–1440.

Llovet, J.M., Bruix, J., Gores, G.J., 2000. Surgical resection versus transplantation for early hepatocellular carcinoma: clues for the best strategy. Hepatology 31, 1019–1021.

Llovet, J.M., Burroughs, A., Bruix, J., 2003. Hepatocellular carcinoma. Lancet 362, 1907–1917.

Llovet, J.M., Fuster, J., Bruix, J., Barcelona-Clinic Liver Cancer Group, 2004. The Barcelona approach: diagnosis, staging, and treatment of hepatocellular carcinoma. Liver Transplant. 10 (2 Suppl. 1), S115–S120.

Llovet, J.M., Schwartz, M., Mazzaferro, V., 2005. Resection and liver transplantation for hepatocellular carcinoma. Semin. Liver Dis. 25, 181–200.

Luo, Z., Xie, Y., Deng, M., Zhou, X., Ruan, B., 2011. Prevalence of hepatitis B in the southeast of China: a population-based study with a large sample size. Eur. J. Gastroenterol. Hepatol. 23, 695.

Luz, P.M., Codeço, C.T., Massad, E., Struchiner, C.J., 2003. Uncertainties regarding dengue modeling in Rio de Janeiro, Brazil. Mem. Inst. Oswaldo Cruz 98, 871–878.

Machado, M.C.C., 1972. Editorial: transplantation of the liver. Rev. Hosp. Clin. Fac. Med. Sao Paulo 27.

Majeed, T.A., Wai, C.T., Rajekar, H., Lee, K.H., Wong, S.Y., Leong, S.O., Singh, R., Tay, K.H., Chen, J., Tan, K.C., 2008. Experience of the transplant team is an important factor for posttransplant survival in patients with hepatocellular carcinoma undergoing living-donor liver transplantation. Transplant. Proc. 40, 2507–2509.

Maksoud, J.G., Chapchap, P., Porta, G., Miura, I., Carone Filho, E., Tannuri, U., et al. Liver transplantation in children: initial experience of the Instituto da Criança of the.

Malinchoc, M., Kamath, P.S., Gordon, F.D., Peine, C.J., Rank, J., ter Borg, P.C., 2000. A model to predict poor survival in patients undergoing transjugular intrahepatic portosystemic shunts. Hepatology 31 (4), 864–871.

Mallick, I.H., Yang, W., Winslet, M.C., Seifalian, A.M., 2004. Ischemia-reperfusioninjury of the intestine and protective strategies against injury. Dig. Dis. Sci. 49, 1359.

Marcen, R., Orfino, L., Pascual, J., 1998. Delayed graft function does not reduce the survival of renal transplant allografts. Transplantation 66, 461–466.

Marsh, J.W., Dvorchik, I., Bonham, C.A., Iwatsuki, S., 2000. Is the pathologic TNM staging system for patients with hepatoma predictive of outcome? Cancer 10, 538–543.

Massad, E., 2008. The elimination of Chagas' disease from Brazil. Epidemiol. Infect. 136 (9), 1153–1164.

Massad, E., Burattini, M.N., Ortega, N.R.S., 1999. Fuzzy logic and measles vaccination: designing a control strategy. Int. J. Epidemiol. 10, 550–557.

Massad, E., Coutinho, F.A.B., Burattini, M.N., Lopez, L.F., Struchiner, C.J., 2005. Yellow fever vaccination: how much is enough? Vaccine 10, 3908–3914.

Massad, E., Coutinho, F.A.B., Chaib, E., Burattini, M.N., 2008a. Cost-effectiveness analysis of a hypothetical hepatitis C vaccine as compared to antiviral therapy. Epidemiol. Infect.

Massad, E., Ortega, N.R.S., DeBarros, L.C., Struchiner, C.J., 2008b. Fuzzy Logic in Action: Applications in Epidemiology and Beyond. Springer-Verlag, Heildeberg.

Massad, E., Coutinho, F.A.B., Chaib, E., Burattini, M.N., 2009. Cost-effectiveness analysis of a hypothetical hepatitis C vaccine as compared to antiviral therapy. Epidemiol. Infect. 137 (2), 241–249.

Matsumo, N., Sakurai, E., Tamaki, I., 1994. Effectiveness of machine perfusion preservation as a viability determination method for kidneys procured from non-heart-beating donors. Transplant. Proc. 26, 2421.

Matsuno, N., Sakurai, E., Uchiyama, M., 1998. Role of machine perfusion preservation in kidney transplantation from non-heart-beating donors. Clin. Transpl. 12, 1–4.

Mayr, E., 1982. The Growth of Biological Thought. The Belknap Press of Harvard University Press.

Mazza, G., Al-Akkad, W., Rombouts, K., Pinzani, M., 2017. Liver tissue engineering:from implantable tissue to whole organ engineering. HepatolCommun 2, 131–141.

Mazzaferro, V., Regalia, E., Doci, R., Andreola, S., Pulvirenti, A., Bozzetti, A., Montalto, F., Ammatuna, M., Morabito, A., Gennari, L., 1996. Liver transplantation for the treatment of small hepatocellular carcinomas in patients with cirrhosis. N. Engl. J. Med. 334, 693–699.

Mazzaferro, V., Llovet, J.M., Miceli, R., Bhoori, S., Schiavo, M., Mariani, L., Camerini, T., Roayaie, S., Schawartz, M.E., Grazi, G.L., Adam, R., Neuhaus, P., Salizzoni, M., Bruix, J., Forner, A., De Carlis, L., Cillo, U., Burroughs, A., Troisi, R., Rossi, M., Gerunda, G.E., Lerut, J., Belghiti, J., Boin, I., Gugenheim, J., Rochling, F., Van Hoek, B., Majno, P., 2009. Metroticket investigator Study Group. Predicting survival after liver transplantation in patients with hepatocellular carcinoma beyond Milan criteria: a retrospective exploratory analysis. Lancet Oncol. 10, 35–43.

Metcalfe, M.S., Nicholson, M.L., 2000. Non-heart-beating donors for renal transplantation. Lancet 356, 1853.

Metcalfe, M.S., Butterworth, P.C., White, S.A., Saunders, R.N., Murphy, G.J., Taub, N., et al., 2001. A case-control comparison of the results of renal transplantation from heart-beating and non-heart-beating donors. Transplantation 71, 1556–1559.

Mies, S., Massarollo, P.C., Baia, C.E., et al., 1988. Liver transplantation in Brazil. Transplant. Proc. 30, 2880.

Mohanty, S., Sanger, K., Heiskanen, A., Trifol, J., Szabo, P., Dufva, M., et al., 2016. Fabrication of scalable tissue engineering scaffolds with dual-pore microarchitecture by combining 3D printing and particle leaching. Mater. Sci. Eng. C 61, 180–189.

Morgan, R.L., Baack, B., Smith, B.D., Yartel, A., Pitasi, M., Falck-Ytter, Y., 2013. Eradication of hepatitis C virus infection and the development of hepatocellular carcinoma: a meta-analysis of observational studies. Ann. Intern. Med. 158, 329–337.

Moya, A., Berenguer, M., Aguilera, V., Juan, F.S., Nicolás, D., Pastor, M., López-Andujar, R., Rayón, M., Orbis, F., Mora, J., De Juan, M., Carrasco, D., Vila, J.J., Prieto, M., Berenguer, J., Mir, J., 2002. Hepatocellular carcinoma: can it be considered a controversial indication for liver trasnplanation in centers with high rates of hepatiotis C? Liver Transplant. 8, 1020–1027.

MSB, 2012. Ministério da Saúde do Brasil, 30 anos do Programa Nacional de Imunizações. Available from: http://portal.saude.gov.br/portal/saude/profissional/area.cfm?id_area=1448.

Muiesan, P., Jassem, W., Girlanda, R., Steinberg, R., Vilca-Melendez, H., Mieli-Vergani, G., et al., 2006. Segmental liver transplantation from non-heartbeating donors- an early experience with implications for the future. Am. J. Transplant. 6, 1012–1016.

Nadalin, S., Bockhorn, M., Malago, M., et al., 2006. Living donor liver transplantation. HPB (Oxford) 8, 10.

Nagai, S., Chau, L.C., Schilke, R.E., Safwan, M., Rizzari, M., Collins, K., Yoshida, A., Abouljoud, M.S., Moonka, D., 2018. Effects of allocating livers for transplantation based on model for end-stage liver disease–sodium scores on patient outcomes. Gastroenterology 155, 1451–1462.

Navarro, A.P., Sohrabi, S., Wilson, C., Sanni, A., Wyrley-Birch, H., Vilayanand, D., et al., 2006a. Renal transplants from category III non-heart-beating donors with evidence of pre-arrest acute renal failure. Transplant. Proc. 38, 2635–2636.

Navarro, A.P., Sohri, S., Wyrley-Birch, H., Vilayanand, D., Wilson, C., Sanni, A., et al., 2006b. Dual renal transplantation from kidneys from marginal non-heart-beating donors. Transplant. Proc. 38, 2633–2634.

Neto, J.S., Carone, E., Pugliese, V., Salzedas, A., Fonseca, E.A., Teng, H., et al., 2007. Living donor liver transplantation for children in Brazil weighing less than 10 kilograms. Liver Transplant. 13, 1153–1158.

Neuberger, J., Gimson, A., Davies, M., Akyol, M., O'Grady, J., Burroughs, A., Hudson, M., 2008. Selection of patients for liver transplantation and allocation of donated livers in the UK. Gut 57, 252–257.

Neuhaus, P., Jonas, S., Bechstein, W.O., Wex, C., Kling, N., Settmacher, U., al-Abadi, H., 1999. Liver transplantation for hepatocellular carcinoma. Transplant. Proc. 31, 469–471.

Newman, C.P., Baxby, K., Hall, R., Taylor, R.M., 1975. Letter machine-perfused cadaver kidneys. Lancet 2, 614.

Nicholson, M.L., 1996. Kidney transplantation from asystolic donors. Br. J. Hosp. Med. 55, 51.

Nicholson, M.L., 2000. Renal transplantation from non-heart-beating donors: opportunities and challenges. Transplant. Rev. 14, 1–17.

Nicholson, M.L., Wheatley, T.J., Horsburgh, T., Edwards, C.M., Veitch, P.S., Bell, P.R.F., 1996. The relative influence of delayed graft function and acute rejection on renal transplant survival. Transpl. Int. 9, 415–419.

Nicholson, M.L., Horsburgh, T., Doughman, T.M., Wheatley, T.J., Butterworth, P.C., Veitch, P.S., 1997. Comparison of the results of renal transplants from conventional and non-heart-beating cadaveric donors. Transplant. Proc. 29, 1386–1387.

Nicholson, M.L., Metcalfe, M.S., White, S.A., Waller, J.R., Doughman, T.M., Horsburgh, T., et al., 2000. A comparison of the results of renal transplantation from non-heart-beating, conventional cadaveric, and living donors. Kidney Int. 58, 2585–2591.

Olson, L., Castro, V.L., Ciancio, G., 1996. Twelve years experience with non-heart-beating cadaveric donors. J. Transpl. Coord. 6, 196.

Onaca, N., Davis, G.L., Goldstein, R.M., Jennings, L.W., Klintmalm, G.B., 2007. Expanded criteria for liver transplantation in patients with hepatocellular carcinoma: a report from the International Registry of Hepatic Tumors in Liver Transplantation. Liver Transplant. 10, 391–399.

OPTN. Donation & Transplantation – About Transplantation: Transplant Process. [online]. USA. Organ Procurement and Transplantation Network. Health Resources and Services Administration. Available at: http://optn.transplant.hrsa.gov/about/transplantation/transplantProcess.asp. [Accessed March 2010].

Ortega, N., Barros, L.C., Massad, E., 2003. Fuzzy gradual rules in epidemiology. Kybernetes 10, 460–477.

Ortiz, A.M., Troncoso, P., Kahan, B.D., 2003. Prevention of renal ischemic reperfusion injury using FTY 720 and ICAM-1 antisense oligonucleotides. Transplant. Proc. 35, 1571.

Pérez, E.V., Castroagudin, J.F., 2010. The future of liver transplantation. Transplant. Proc. 42, 613–616.

Petruzziello, A., Marigliano, S., Loquercio, G., Cacciapuoti, C., 2016. Hepatitis C virus (HCV) genotypes distribution: an epidemiological up-date in Europe. Infect. Agents Cancer 11, 53.

Pfaff, W.W., Howard, R.J., Patton, P.R., Adams, V.R., Rosen, C.B., Reed, A.L., 1998. Delayed graft function after renal transplantation. Transplantation 65, 219–223.

Raia, S., Nery, J.R., Mies, S., 1989. Liver transplantation from live donos. Lancet 2, 497.

Rice, H.E., O'Keefe, G.E., Helton, W.S., et al., 1997. Morbid prognostic features in patients with chronic liver failure undergoing nonhepatic surgery. Arch. Surg. 132, 880–884.

Ringe, B., Pichlmayr, R., Wittekind, C., Tusch, G., 1991. Surgical treatment of hepatocellular carcinoma: experience with liver resection and transplantation in 198 patients. World J. Surg. 10, 270–285.

Roayaie, S., Frischer, J.S., Emre, S.H., Fishbein, T.M., Sheiner, P.A., Sung, M., Miller, C.M., Schwartz, M.E., 2002. Long-term results with multimodal adjuvant therapy and liver transplantation for the treatment of hepatocellular carcinomas larger than 5 centimeters. Ann. Surg. 10, 533–539.

Roayaie, S., Schwartz, J.D., Sung, M.W., Emre, S.H., Miller, C.M., Gondolesi, G.E., Krieger, N.R., Schwartz, M.E., 2004. Recureence of hepatocellular carcinoma after liver transpl:patterns and prognosis. Liver Transplant. 10, 534–540.

Roberts, M.S., Angus, D.C., Bryce, C.L., Valenta, Z., Weissfeld, L., 2004. Survival after liver transplantation in the United States: a disease-specific analysis of the UNOS Database. Liver Transplant. 10, 886–897.

Rode, A., Bancel, B., Douek, P., Chevallier, M., Vilgrain, V., Picaud, G., Henry, L., Berger, F., Bizollon, T., Gaudin, J.L., Ducerf, C., 2001. Small nodule detection in cirrhotic livers: evaluation with US, spiral CT and MRI and correlation with pathologic examination of explanted liver. J. Comput. Assist. Tomogr. 25, 327–336.

Royal Colleges, 1996. Conference of Medical Royal Colleges and their faculties in the United Kingdom. Diagnosis of brain stem death. Lancet ii, 1069–1070.

Ryckman, F.C., Bucuvalas, J.C., Nathan, J., Alonso, M., Tiao, G., Balistreri, W.F., 2008. Outcomes following liver transplantation. Semin. Pediatr. Surg. 10, 123–130.

Samuel, D., Colombo, M., El-Serag, H., Sobesky, R., Heaton, N., 2011. Toward optimizing the indications for orthotopic liver transplantation in hepatocellular carcinoma. Liver Transplant. 10. S6–S13.

Santoyo, J., Suarez, M.A., Fernádez-Aguilar, J.L., Jiménez, M., Perez Daga, J.A., Sánchez-Perez, B., Gonzalez Poveda, I., Gonzalez-Sanchez, A., Ramírez, C., de la Fuente, A., 2005. Liver transplant results for hepatocellular carcinoma applying strict preoperative selection criteria. Transplant. Proc. 37, 1488–1490.

Sauer, P., Kraus, T.W., Schemmer, P., Mehrabi, A., Stremmel, W., Buechler, M.W., Encke, J., 2005. Liver transplantation for hepatocellular carcinoma: is there evidence for expanding the selection criteria? Transplantation 80 (1 Suppl.), S105–S108.

Schauer, R.J., Kalmuk, S., Gerbes, A.L., Leiderer, R., Meissner, H., Schildberg, F.W., 2004. Intravenous administration of glutathione protects parenchymal and non-parenchymal liver cells against reperfusion injury following rat liver transplantation. World J. Gastroenterol. 10, 864.

Schlumpf, R., Weber, M., Weinreich, T., Spahn, D., Rothlin, M., Candinas, D., 1996. Transplantation of kidneys from non-heart-beating donors: protocol, cardiac death diagnosis and results. Transplant. Proc. 28, 107.

Schwartz, R., Dameshek, W., 1959. Drug induced immunologic tolerance. Nature 183, 1682–1683.

Scott, D.F., Morley, A.R., Swinney, J., 1969. Canine renal preservation following hypothermic perfusion and subsequent function. Br. J. Surg. 56, 688.

Scwartz, M., 2004. Liver transplantation in patients with hepatocellular carcinoma. Liver Transplant. 10 (2 Suppl. 1), S81–S85.

Selzner, N., Rudiger, H., Graf, R., Clavien, P., 2003. Protective strategies against ischemic injury of the liver. Gastroenterology 125, 917.

Shiroki, R., Hoshinaga, K., Higuchi, T., 1998. Prolonged warm ischemia affects long-term prognosis of kidney transplant allografts from non-heartbeating donors. Transplant. Proc. 30, 111.

Silva, M., Moya, A., Berenguer, M., Sanjuan, F., López-Andujar, R., Pareja, E., Torres-Quevedo, R., Aguilera, V., Montalva, E., De Juan, M., Mattos, A., Prieto, M., Mir, J., 2008. Expanded criteria for liver transplanation in patients with cirrhosis and hepatocellular carcinoma. Liver Transplant. 14, 1449–1460.

Soderstrom, T., Stoica, P., 1989. System Identification Prentice Hall International.

Soejima, Y., Yoshizumi, T., Uchiyama, H., Aishima, S., Terashi, T., Shimada, M., Maehara, Y., 2007. Extended indication for living donor liver transplantation in patients with hepatocellular carcinoma. Transplantation 83, 893–899.

Sohrabi, S., Navarro, A., Wilson, C., Sanni, A., Wyrley-Birch, H., Anand, V., et al., 2006a. Diabetic donors as a source of non-heart-beating renal transplants. Transplant. Proc. 38, 3402–3403.

Sohrabi, S., Navarro, A., Wilson, C., Asher, J., Sanni, A., Wyrley-Birch, H., et al., 2006b. Renal graft function after prolonged agonal time in non-heartbeating donors. Transplant. Proc. 38, 3400–3401.

Starzl, T.E., Marchiori, T.L., Waddell, W.R., 1963a. The reversal rejection in human renal homografts with subsequent development of homograft tolerance. Surg. Gynecol. Obstet. 117, 385–395.

Starzl, T.E., Marchioro, T.L., Von Kaulla, K., Herrmann, G., Brittain, R.S., Wadell, W.R., 1963b. Homotransplantation of the liver in humans. Surg. Gynecol. Obstet. 117, 659.

Starzl, T.E., Klintmalm, G.B.G., Porter, K.A., Iwatsuki, S., SchroTer, G.P., 1981. Liver transplantation with use of cyclosporine-A and prednisone. N. Engl. J. Med. 305, 266–269.

Steingruber, I.E., Mallouhi, A., Czermak, B.V., Waldenberger, P., Gassner, E., Offner, F., Chemelli, A., Koenigsrainer, A., Vogel, W., Jaschke, W.R., 2003. Pretransplantation evaluation of the cirrhotic liver with explantation correlation: accuracy of CT arterioportography and digital subtraction hepatic angiography in revealing hepatocellular carcinoma. Am. J. Roentgenol. 181, 99–108.

Stell, D.A., McAlister, V.C., Thorburn, D., 2004. A comparison of disease severity and survival rates after liver transplantation in the United Kingdom, Canada, and the United States. Liver Transplant. 10, 898–902.

Stoker, J., Romijn, M.G., de Man, R.A., Brouwer, J.T., Weverling, G.J., van Muiswinkel, J.M., Zondervan, P.E., Laméris, J.S., Ijzermans, J.N., 2002. Prospective comparative study of spiral computer tomography and magnetic resonance imaging for detection of hepatocellular carcinoma. Gut 51, 105–107.

Stoll-Keller, F., Barth, H., Fafi-Kremer, S., Zeisel, M.B., Baumert, T.F., 2009. Development of Hepatitis C virus vaccines: challenges and progress. Expert Rev. Vaccines 8 (3), 333–345.

Sudhindran, S., Pettigrew, G.J., Drain, A., Shroti, M., Watson, C.J.E., Jamieson, N.V., et al., 2003. Outcome of transplantation using kidneys from controlled (Maastrich category 3) non-heart-beating donors. Clin. Transplant. 17, 93–100.

Takada, Y., Ito, T., Ueda, M., Sakamoto, S., Haga, H., Maetani, Y., Ogawa, K., Ogura, Y., Oike, F., Egawa, H., Uemoto, S., 2007. Living donor liver transplantation for patients with HCC exceeding the Milan criteria: a proposal of expanded criteria. Dig. Dis. 25, 299–302.

Tanabe, K., Oshima, T., Tokumoto, T., 1998. Long-term renal function in on non-heart-beating donor kidney transplantation: a single centre experience. Transplantation 66, 1708.

Tanwar, S., Khan, S.A., Grover, V.P., Gwilt, C., Smith, B., Brown, A., 2009. Liver transplantation for hepatocellular carcinoma. World J. Gastroenterol. 15 (44), 5511–5516.

Teefy, S.A., Hildeboldt, C.C., Dehdashti, F., Siegel, B.A., Peters, M.G., Heiken, J.P., Brown, J.J., McFarland, E.G., Middleton, W.D., Balfe, D.M., Ritter, J.H., 2003. Detection of primary hepatic malignancy in liver candidates: prospective comparison of CT, MR imaging, US and PET. Radiology 226, 533–542.

Terasaki, P.I., Cho, Y.M., Cheka, J.M., 1997. Strategy for eliminating the kidney shortage. Clin. Transpl. 267, 132.

Tesi, R., Elkammas, E.A., Davies, E.A., 1993. Pulsatile kidney perfusion and evaluation: use of high risk kidney donors to expand the donor pool. Transplant. Proc. 25, 3099–4100.

Testa, G., Vidanovic, V., Chejfec, G., et al., 2008. Adult living-donor liver transplantation with ABO-incompatible grafts. Transplantation 85, 681.

Thabut, G., Brugiere, O., Leseche, G., Stern, J.B., Fradj, K., Herve, P., 2001. Preventive effectof inhalet nitric oxide and pentoxifylline on ischemia/perfusion injury after lung transplantation. Transplantation 71, 1295.

Thuluvath, P.J., Guidinger, M.K., Fung, J.J., Johnson, L.B., Rayhill, S.C., Pelletier, S.J., 2010. Liver transplantation in the United States, 1999–2008. Am. J. Transplant. 10, 1003–1019.

Todo, S., Furukawa, H., Tada, M., 2007. Japanese Liver Transplantation Study Group. Extending indication: role of living donor liver transplantation for hepatocellular carcinoma. Liver Transplant. 13 (11 Suppl. 2), S48–S54.

Tosa, C., Dupuis-Lorezon, E., Majno, P., Berney, T., Kneteman, N.M., Perneger, T., Morel, P., Mentha, G., Combescure, C., 2012. A model for dropout assessment of candidates with or without hepatocellular carcinoma on a common liver transplant waiting list. Hepatology 10, 149–156.

Troppman, C., Gillinham, K.J., Gruesnerr, W.G., 1996. Delayed graft function in the absence of rejection has no long-term impact. A study of cadaver kidneys recipients with good graft function at 1 year after transplantation. Transplantation 61, 1331–1337.

UK Transplant, 2004. Transplant Activity in the UK 2003–2004. Bristol: UK transplant. More transplants, New Lives.

UK Transplant Activity, 2001. htpp://www.uktransplant.org.uk/statistics/transplant_activity/uk_trans_activity 2001.

UK Transplant Activity, 2005/2006. Available at: htpp://www.uktransplant.org.uk/statistic/transplant_activity/uk_trans_activity2006. [Accessed February 2008].

UNOS. Liver Policy Outcomes Encouraging at Six Months; Some Adjustments Recommended. [online]. USA. United Network for Organ Sharing. Available at: http://www.optn.org/news/newsDetail.asp?id=227. [Accessed March 2010].

Van der Vliet, J.A., Vroemen, P.A.M., Cohen, B., Kootstra, G., 1984. Comparison of cadaver kidney preservation methods in Eurotransplant. Transplant. Proc. 16, 180–181.

Van der Werf, W.J., D'Alessandro, A.M., Hoffmann, R.M., 1988. Procurement, preservation, and transport of cadaver kidneys. Surg. Clin. N. Am. 78, 570–572.

Varty, K., Veitch, P.S., Morgan, J.D.T., Bell, P.R.F., 1994. Kidney retrieval from asystolic donors: a valuable and viable source of additional organs. Br. J. Surg. 81, 1459.

Voigt, M.D., Zimmerman, B., Katz, D.A., Rayhill, S.C., 2004. New national liver transplant allocation policy: is the regional review board process fair? Liver Transplant. 10 (5), 666–674.

Volk, M.L., Vijan, S., Marrero, J.A., 2008. A novel model measuring the harm of transplanting hepatocellular carcinoma exceeding Milan criteria. Am. J. Transplant. 10, 839–846.

Weber, M., Dindo, D., Demartines, N., Ambuhl, P.M., Clavien, P.A., 2002. Kidney trasplantation from donors without a heartbeat. N. Engl. J. Med. 347, 248.

Wheatley, T.J., Doughman, T.M., Veitch, P.S., Nicholson, M.L., 1996. Kidney retrieval from asystolic donors using an intra-aortic balloon catheter. Br. J. Surg. 83, 962–963.

WHO. Death Rate Trends for RTAs and CVAs. WHO European Health for All database – http://www.euro.who.int/hfadb. [Accessed 28 July 2010].

Wiesner, R., Edwards, E., Freeman, R., Harper, A., Kim, R., Kamath, P., Kremers, W., Lake, J., Howard, T., Merion, R.M., Wolfe, R.A., Krom, R., United Network for Organ Sharing Liver Disease Severity Score Committee, 2003a. Model for end-stage liver disease (MELD) and allocation of donor livers. Gastroenterology 124, 91–96.

Wiesner, R., Edwards, E., Freeman, R., et al., 2003b. Model for end-stage liver disease (MELD) and allocation of donor livers. Gastroenterology 124 (1), 91–96.

Wiesner, R., Lake, J.R., Freeman Jr., R.B., Gish, R.G., 2006. Model for end-stage liver disease (MELD) exception guidelines. Liver Transplant. 12 (Suppl. 3), S85–S87.

Wijnen, R.M., Booster, M.H., Stubenitsky, B.M., de Boer, J., Heineman, E., Kootstra, G., 1995. Outcome of transplantation of non-heart-beating donors kidneys. Lancet 345, 1067.

Wilson, E.O., 1998. Concilience, the Unity of Knowledge. Knopf, Nova Iork, EUA.

Wilson, C.H., Brook, N.R., Gok, M.A., Asher, J.F., Nicholson, L., Talbot, D., 2005. Randomized clinical trial of daclizumab induction and delayed introduction of tracolimus recipients of non-heart-beating kidney transplants. Br. J. Surg. 92, 681–687.

Wojcicki, M., Lubikowski, J., Wrzesinski, M., Post, M., Jarosz, K., Hydzik, P., et al., 2007. Outcome of emergency liver transplantation including mortality on the waiting list: a single-center experience. Transplant. Proc. 39, 2781–2784.

Xiao, L., Fu, Z.R., Ding, G.S., Fu, H., Ni, Z.J., Wang, Z.X., Shi, X.M., Guo, W.Y., 2009. Liver transplantation for hepatitis B virus-related hepatocellular carcinoma: one center's experience in China. Transplant. Proc. 41, 1717–1721.

Yao, F.Y., Ferrell, L., Bass, N.M., Watson, J.J., Bacchetti, P., Venook, A., Ascher, N.L., Roberts, J.P., 2001a. Liver transplantation for hepatocellular carcinoma: expansion of the tumor size limits does not adversely impact survival. Hepatology 33, 1394–1403.

Yao, F.Y., Ferrel, L., Bass, N.M., Watson, J.J., Bachetti, P., Venook, A.L., 2001b. Liver transplantation for hepatocellular carcinoma: expansion of the tumor size limits does not adversely impact survival. Hepatology 10, 1394–1403.

Yao, F.Y., Roberts, J.P., 2004. Applying expanded criteria to liver transplantation for hepatocellular carcinoma: too much too soon, or is now the time. Liver Transplant. 10, 919–921.

Yao, F.Y., Xiao, L., Bass, N.M., Kerlan, R., Ascher, N.R., Roberts, J.P., 2007. Liver transplantationm for hepatocellular carcinoma: validation of the UCSF-enpanded criteria based on preoperative imaging. Am. J. Transplant. 7, 2587–2596.

Yokoyama, I., Uchida, K., Kobayashi, Y., Orihara, A., Takagi, H., 1994. Effect of prolonged delays graft function on long-term graft outcome in cadaveric kidney transplantation. Clin. Transplant. 8, 101.

Yokoyama, I., Uchida, K., Hayashi, S., 1996. Factors affecting graft function in cadaveric renal transplantation from non-heart-beating donors using a double balloon catheter. Transplant. Proc. 28, 116.

Zavaglia, C., De Carlis, L., Alberti, A.B., Minola, E., Belli, L.S., Slim, A.O., Airoldi, A., Giacomoni, A., Rondinara, G., Tinelli, C., Forti, D., Pinzello, G., 2005. Predictors of longterm survival after liver transplantation for hepatocellular carcinoma. Am. J. Gastroenterol. 100, 2708–2716.
Zhang, H., Li, Q., Sun, J., Wang, C., Gu, Q., Feng, X., 2011. Seroprevalence and risk factors for hepatitis B infection in an adult population in Northeast China. Int. J. Med. Sci. 8, 321.

Further reading

Alvarez-Rodrigues, J., Barrio, R., Martin, M., 1997. Factors influencing short and long-term survival of kidneys transplanted from non-heart-beating donors. Transplant. Proc. 29, 3490.
Bacchella, T., Machado, M.C.C., 2004. The first clinical liver transplantation of Brazil revisited. Transplant. Proc. 36, 929–930.
Belzer, F.O., Southard, J.H., D'Alessandro, A.M., Sollinger, H.W., Kalayoglu, M., 1993. Update on preservation of liver grafts. Transplant. Proc. 25, 2010–2011.
Bertevello, P.L., Chaib, E., 2002. Hepatic artery system variations correlated to split-liver surgery: anatomic study in cadavers. Arq. Gastroenterol. 39, 81–85.
Boyer, N., Marcellin, P., 2000. Natural history of hepatitis C and the impact of anti-viral therapy. Forum (Genova) 10 (1), 4–18.
Brown Jr., R.S., 2008. Live donors in liver transplantation. Gastroenterology 134, 1802.
Bruix, J., Llovet, J.M., 2002. Prognostic prediction and treatment strategy in hepatocellular carcinoma. Hepatology 35, 519–524.
Chaib, E., 1993. Transplante de fígado: alterações da artéria hepática e do fígado em 80 doadores. Arq. Gastroenterol. 30, 82–87.
Collins, G.M., Bravo-Shugarman, M., Terasaki, P.I., 1969. kidney preservation for transportation:initial perfusion and 30 hours ice storage. Lancet 2, 1219.
Coombes, J.M., James, F., Trotter, J.F., 2005. Development of the allocation system for deceased donor liver transplantation. Clin. Med. Res. 3 (2), 87–92.
Ferraz, A.S., Pereira, L.A., Correa, M.C., 1999. Transplants with cadaveric organs in the State of São Paulo, Brazil: the new organizational model and first results. Transplant. Proc. 31, 3075.
Fung, J., Marsh, W., 2002. The quandary over liver transplantation for hepatocellular carcinoma: the greater sin? Liver Transplant. 8, 775–777.
Gerstenkorn, C., 2003. Non-heart-beating donors: renewed source of organs for renal transplantation during the twenty-first century. World J. Surg. 27, 489–493.
Grünwald, P., 2007. The Minimum Description Length Principle. MIT Press, Cambridge.
Hoshinaga, K., Shiroki, R., Fujita, T., Kanno, T., Naide, Y., 2005. The fate of 359 renal allografts harvested from non-heart-beating cadáver donors at a single center. In: Hwang, S., Lee, S.G., Joh, J.W., Suh, K.S., Kim, D.G. (Eds.), Liver Transplantation for Adult Patients with Hepatocellular Carcinoma in Korea: Comparison between Cadaveric Donor and Living Donor Liver Transplantations, Liver Transpl, vol. 11, pp. 1265–1272.
Hospital das Clínicas of Universidade de São Paulo (in Portuguese). AMB Rev. Assoc. Med. Bras. 37, 1991, 193–199.
Humar, Beissel, J., Crotteau, S., et al., 2008. Whole liver versus split liver versus living donor in the adult recipient: an analysis of outcomes by graft type. Transplantation 85, 1420.
Jamieson, N.V., 1989. A new solution for liver preservation. Br. J. Surg. 76, 107–108.
Jamieson, N.V., 1991. Review article:improved preservation of the liver transplantation. Aliment Pharmacol. Ther. 5, 91–104.
Jamieson, N.V., Sundberg, R., Lindell, S., Southard, J.H., Bezer, F.O., 1988a. An analisys of the components in UW solution using the isolated perfused rabit liver. Transplantation 46, 512–516.

Jamieson, N.V., Sundberg, R., Lindell, S., Laravuso, R., Kalayoglu, M., Southard, J.H., Bezer, F.O., 1988b. Successful 24 to 30 − hour preservation of canine liver: a preliminary report. Transplant. Proc. 20, 945−947.
Jamieson, N.V., Sundberg, R., Lindell, S., Claesson, K., Moen, J., Vreugdenhil, P.K., Wight, D.G.D., Southard, J.H., Bezer, F.O., 1988c. Preservation of the canine liver for 24−48 hours using simple cold storage with UW solution. Transplantation 46, 517−522.
Jawaid, A., Khuwaia, A.K., 2008. Treatment and vaccination for hepatitis C: present and future. J. Ayub Med. Coll. Abbottabad 20 (1), 129−133.
Jochmans, I., van Rosmalen, M., Pirenne, J., Samuel, U., 2017. Adult liver allocation in eurotransplant. Transplantation 101 (7), 1542−1550. https://doi.org/10.1097/TP.0000000000001631.
Johnson, R.W., 1972. The effect of ischaemic injury on kidneys preserved for 24 h before transplantation. Br. J. Surg. 59, 765.
Kandula, P., Anderson, T.A., Vagefi, P.A., 2013. Allocation of resources for organ transplantation. Anesthesiol. Clin. 31 (4), 667−674. https://doi.org/10.1016/j.anclin.2013.08.002.
Klintmalm, G.B., 1998. Liver transplantation for hepatocellular carcinoma: a registry report of the impact of tumor characteristics on outcome. Ann. Surg. 228, 479−490.
Kneteman, N., Oberholzer, J., Al Saghier, M., Meeberg, G.A., Blitz, M., Ma, M.M., Wong, W.W., Gutfreund, K., Mason, A.L., Jewell, L.D., Shapiro, A.M., Bain, V.G., Bigam, D.L., 2004. Sirolimus-based immunosuppression for liver transplantation in the presence of extended criteria for hepatocellular carcinoma. Liver Transplant. 10, 1301−11.
Kozaki, M., Matsuno, N., Tamaki, T., 1991. Procurement of kidney Grafts from non-heart-beating donors. Transplant. Proc. 23, 2575.
Lichtman, S.N., Lemasters, J.J., 1999. Role of cytokines and cytokine-producing cells in reperfusion injury to the liver. Semin. Liver Dis. 19, 171.
Lu, F.M., Zhuang, H., 2009. Prevention of hepatitis B in China: achievements and challenges. Chin. Med. J. (Engl) 122, 2925.
Makisalo, H., Chaib, E., Krokos, N., Calne, R.Y., 1993. Hepatic arterial variations and liver related diseases of 100 consecutive donors. Transpl. Int. 6, 325−329.
Makuuchi, M., Sano, K., 2004. The surgical approach to HCC: our progress and results in Japan. Liver Transplant. 10 (2 Suppl. 1), S46−S52.
Marsh, J.W., Dvorchik, I., 2003. Liver organ allocation for hepatocellular carcinoma: are we sure? Liver Transplant. 9, 693−696.
Matesanz, R., Miranda, B., 1995. Outcome of transplantation of non-heartbeating kidneys. Lancet 346, 53.
McCormick, A., Sultan, J., 2005. Liver transplantation—patient selection and timing. Med. J. Malays. 60 (Suppl. B), 83−87.
Moyer, V.A., U.S. Preventive Services Task Force, 2013. Screening for hepatitis C virus infection in adults: U.S. Preventive Services Task Force recommendation statement. Ann. Intern. Med. 159, 349−357. https://doi.org/10.7326/0003-4819-159-5-201309030-00672.
Murray, J.E., Merril, J.P., Dammin, G.J., 1962. Kidney transplantation in modified recipients. Ann. Surg. 156, 337.
Opelz, G., Terasaki, P.I., 1981. Advantage of cold storage over machine perfusion for preservation of cadaver kidneys. Transplantation 33, 64−68.
Penn, I., 1991. Hepatic transplantation for primary and metastic cancers of the liver. Surgery 110, 726−734.
Prium, J., Klompmaker, I.J., Haagsma, E.B., Bijleveld, C.M.A., 1993. Sloof MJH − selection criteria for liver deonation: a review. Transpl. Int. 6, 226−235.
Reddy, S., Zilvetti, M., Brockmann, J., McLaren, A., Friend, P., 2004. Liver transplantation fron non-heart-beating donors: current status and future prospects. Liver Transplant. 10, 1223−1232.
Rissanen, J., 1986. Stochastic complexity and modeling. Ann. Stat. 14 (3), 1080−1100.

Robertson, J.A., 1988. Relaxing the death standard for organ donation in paediatric situations. In: Mathieu, D. (Ed.), Organ Substitution Technology: Ethical, Legal and Public Policy Issues. Westview, Boulder, p. 69.

Ryder, S.D., British Society of Gastroenterology, 2003. Guidelines for the diagnosis and treatment of hepatocellular carcinoma (HCC) in adults. Gut 52 (Suppl. 3), iii1–iii8.

Schwarz, G., 1978. Estimating the dimension of a model. Ann. Stat. 6 (2), 461–464.

Sharma, P., Balan, V., Hernandez, J.L., Harper, A.M., Edwards, E.B., Rodriguez-Luna, H., Byrne, T., Vargas, H.E., Mulligan, D., Rakela, J., Wiesner, R.H., 2004. Liver transplantation for hepatocellular carcinoma: the MELD impact. Liver Transplant. 10, 36–41.

Starzl, T.E., Koep, L.J., Halgrimson, C.G., et al., 1979. Fifteen years of clinical transplantation. Gastroenterology 77, 375–388.

Takada, Y., Uemoto, S., 2010. Liver transplantation for hepatocellular carcinoma: the Kyoto experience. J. Hepatobiliary Pancreat. Sci. 17, 527–532.

Tan, C.K., Gores, G.J., Steers, J.L., Porayko, M.K., Hay, J.E., Rakela, J., Wiesner, R.H., Krom, R.A., 1994. Orthotopic liver transplantation for preoperative early-stage hepatocellular carcinoma. Mayo Clin. Proc. 69, 509–514.

Terasaki, P.I., Cecka, J.M. (Eds.), 1998. Clinical Transplantation. UCLA Tissue Typing Laboratory, Los Angeles.

The British Transplantation Society, 1998. Towards Standards for Organ and Tissue Transplantation in the United Kingdom. BTS, Richmond.

Toniutto, P., Zanetto, A., Ferrarese, A., Burra, P., 2017. Current challenges and future directions for liver transplantation. Liver Int. 37, 317–327.

United Network for Organ Sharing Donation & Transplantation – Policy 3.6. Allocation of Livers. [online]. USA. Organ Procurement and Transplantation Network. Health Resources and Services Administration. Available at: http://www.optn.org/organDatasource/OrganSpecificPolicies.asp?display=Liver. [Accessed March 2010].

United Network for Organ Sharing, 2004. UNOS Annual Report: Based on Organ Procurement and Transplantation Network Data 2005. UNOS, Richmond, VA.

Yao, F.Y., 2007. Expanded criteria for liver transplantation in patients with hepatocellular carcinoma. Hepatol. Res. 37 (Suppl. 2), S267–S274.

Yu, M.L., Chuang, W.L., 2009. Treatment of chronic hepatitis C in Asia: when East meets west. J. Gastroenterol. Hepatol. 24 (3), 336–345.

Index

Note: 'Page numbers followed by "*f*" indicate figures and "*t*" indicate tables'.

H

I

J

K

L

M

Printed in the United States
By Bookmasters